POLITEXT 177

Technical Writing

A Guide for Effective Communication

POLITEXT

Carmen Bombardó Solés
Marta Aguilar Pérez
Clàudia Barahona Fuentes

Technical Writing
A Guide for Effective Communication

EDICIONS UPC

Primera edició: setembre de 2007
Segona edició: setembre de 2008
Reimpresio: julio de 2009

Disseny de la coberta: Manuel Andreu

©   els autors, 2007

©   Edicions UPC, 2007
      Edicions de la Universitat Politècnica de Catalunya, SL
      Jordi Girona Salgado 1-3, 08034 Barcelona
      Tel.: 934 137 540   Fax: 934 137 541
      Edicions Virtuals: www.edicionsupc.es
      E-mail: edicions-upc@upc.edu

Producció: LIGHTNING SOURCE

Dipòsit legal: B-45088-2007
ISBN: 978-84-8301-966-5

For our families

Contents

Preface

1. Purpose and approach

This book arose from the need to have a textbook to teach technical writing to Spanish engineering students at university. Although there are good books on technical writing as well as on writing in general on the market, we could not find one that suited our students' needs: they were either too theoretical, too practical or simply aimed at students with a different cultural background. Because of this, we decided to write a book that kept the balance between the theoretical explanations necessary to understand the basic concepts on which technical writing is based and the practical aspects that would enable students to put into practice these theoretical concepts. After all, writing is a communication skill that is mostly learnt and mastered by practising it; it is not enough to have a good command of grammar and punctuation. Writing is a much more complex task requiring other techniques, such as organizing ideas logically and clearly, joining sentences coherently, using the appropriate tone and style, etc. Mastering writing, as any experienced writer knows, takes time and practice and a good way to improve writing that cannot be overlooked is reading. The more you read, the better you will write since reading is a practice that rubs off by improving vocabulary, grammar and writing techniques in general. Thus, students are encouraged to read as much as possible and from almost any kind of reading—magazines, fiction books, newspapers, novels, Internet articles, etc. Although the kinds of documents technical students will be asked to write may not require the same level of subtlety as for example a novel, they still call for some of the same skills.

The book's approach to writing is integrative and results from drawing on knowledge of three different approaches—product, process and genre. The process approach is the central one to which the other two are subsumed. We took the process approach as the core or central one because we believe it highly contributes to the development of students' writing abilities as it gives much importance to the skills or stages involved in writing. In a word, untrained writers, like most of our students, welcome having some sort of guidance to help them get started and organize their ideas, and this approach has proved to serve this purpose. We should emphasize that this approach is by no means prescriptive but acts more as guidance allowing enough room for manoeuvre so that writers can adapt it to their own writing preferences. However, insofar as the process approach does not cater for certain fundamental teaching aspects such as the linguistic input and the different kinds of texts, it became necessary to incorporate other approaches. In this sense, the product approach accounts for the linguistic knowledge of texts, basically grammar and text structure, and recognizes the importance of the text as a final product. In addition, the genre approach acknowledges that writing takes place in a social context as a response to a particular need and so heeds the writing conventions established by the technical and scientific community. All in all, our approach seeks to merge the linguistic, procedural and social-cultural aspects that intervene in the process of writing technical documents.

2. Book organization

This book has been organized into three main parts:

- Part I. Introduction to Technical Writing
- Part II. The Writing Process
- Part III. Handbook

Part I, as its name indicates, introduces the basic concepts of technical writing: its importance, definition and main characteristics, as well as a brief description of the main functions found in this register.

Part II focuses on the process of writing. Its three main stages—pre-writing, writing and post-writing—are fully developed. Substages and their associated linguistic and structural aspects are also studied in detail.

Part III complements the former two by providing a summary of some language-related aspects such as the main linguistic constituents and punctuation rules. Besides, it includes further practice on the grammatical and stylistic points seen in Part II.

The parts are internally organized into an introduction and one or more chapters. The introduction is aimed at contextualizing and unifying the content of the part. Likewise, the chapters begin with a somewhat theoretical explanation of the topic in question which is complemented with illustrative examples and, whenever possible, visual information, thus facilitating understanding and appeal.

Finally, we would like to highlight the fact that this textbook is best complemented by a good array of the most common technical documents that engineers and technical professionals need to write at the workplace. For reasons of space and length, we decided to devote this textbook to the writing process only, but students should, either simultaneously or after Part II (or Part III), be exposed to a wide range of texts and have extensive and intensive practice in writing all kinds of technical documents for different purposes and situations.

3. Methodology

Because this book is intended to be a practical and useful manual, the theoretical explanations are followed by a wide range of tasks. In addition, in order to meet our students' needs, we have selected a variety of authentic texts from different sources (e.g. textbooks, research articles, magazines, lab manuals, Internet web pages and even newspaper articles) so as to cater for diverse engineering specialities. Having a considerable amount of tasks and texts is a valuable resource for teachers as it allows them to select those they find most convenient for their students. Keeping in mind that the activities are fundamentally task-based, combined with some problem-solution ones, the criterion adopted to organize them was to group the tasks according to whether they could be done individually or collaboratively. We should point out that this task division is not fixed or closed in the sense that individual tasks are only meant to be done individually and collaborative tasks only collaboratively. On the contrary, this classification is quite flexible as teachers can decide how the task can be carried out in class. Taking this into account, the different types of tasks included in this book fall into the following categories:

- *Reflecting on questions*. All chapters begin with what we called a *reflecting on* activity whose main function is to make students aware of different aspects that will be dealt with within the chapter. These awareness-raising questions also anticipate what the chapter is about.

- *Task-based activities*. These tasks aim to make students work with the different writing techniques previously explained within the chapter. Their increasing level of difficulty allows students to gradually become skilled at these techniques. This way, students acquire the different skills necessary to succeed in the more global and authentic problem-and-solution tasks.

- *Problem-and-solution tasks*. With these types of tasks technical students will be trained to work under similar circumstances to those they will find themselves in their future professional career.

- *Critical thinking tasks*. At some key points, evaluative thinking tasks have been included to make students critically analyze different topics and situations. These tasks go beyond subject-matter considerations and allow students to identify weaknesses, assess alternatives and evaluate evidence by making reasoned judgements.

- *Project*. This globalizing activity is divided into three main parts corresponding to the three main stages of the writing process and builds on the tasks within each stage. This project can be carried out according to two main approaches, namely top-down or bottom-up, so that students can choose the option that better suits their idiosyncratic learning style.

Both the flexibility of the tasks and the methodology described above enable this book to be used within the incoming European Educational System as it caters for students' individual needs and learning styles and promotes collaborative learning (which allows for teamwork with assignment of roles) and project work. Besides, the key to the exercises allows for great flexibility and dynamism because teachers can decide which tasks are to be done and corrected in class or at home, peer-reviewed or teacher-reviewed. In a word, the book can also be used as a kind of self-study book, where students become more responsible for their learning process by actively monitoring it.

The book can be used with both undergraduate and graduate students. Undergraduate students who have not received any instruction on technical communication will probably need to carefully read the theoretical explanations preceding most chapters and sections. In contrast, more mature students will either skip or merely glimpse at the introductory framework on their own at home and go straight to the tasks that will help them improve those skills in which they might be less proficient.

Finally, the materials also adapt to teachers with different teaching styles and with different degrees of experience in written communication. For example, teachers with little experience in written communication may well appreciate a structured and reasoned theoretical explanation before plunging into the tasks, whereas more experienced teachers can exploit this theory as a critical thinking or reflecting on task, thus making lessons more dynamic.

Acknowledgements

We thank the authors and publishers of the material cited in this book for kindly giving us reprint permission. Although every effort has been made to contact authors and publishers, this has not always been possible so any information from them will be welcome and omissions or errors will be corrected.

To our students we owe their kind permission to use their written work and their comments because they greatly contributed to a better version of this book.

Thanks are also due to Brian Tomlinson and Hitomi Masuhara from Leeds Metropolitan University for their encouragement and their endless suggestions on innovative possibilities. They made us realize that this book is just a first attempt that will certainly need future revisions, as a textbook can never be a finished product. We are also indebted to Helen East from the Language Unit at the University of Cambridge for her wise and discerning suggestions.

PART I

INTRODUCTION TO TECHNICAL WRITING

CHAPTER 1
What is technical writing?

Technical writing at university

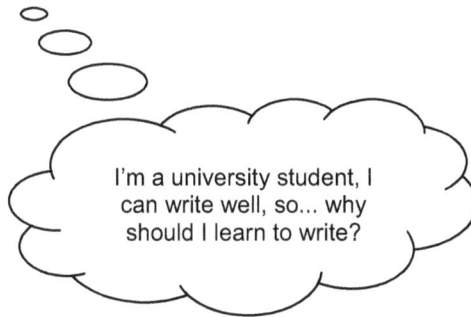

> I'm a university student, I can write well, so... why should I learn to write?

Haven't you ever had this thought, or a similar one? Of course, most of you are more or less competent writers in your first language—maybe some even in English. In fact, many people can get through their lives with just a first language literacy to write postcards, recipes, shopping lists, or odd messages. These documents are quite spontaneous and transient and therefore do not require a large amount of planning. But we are not addressing you as apprentice writers or as English language beginners. Not even as proficient writers in general English. We are addressing you as future skilled professionals who need to perfect their writing skills in English from a professional point of view. This implies that you will need to be acquainted with certain types of documents, known as *genres*, which have specific characteristics (e.g. layout, content or style). The examples mentioned above (a recipe, a postcard or the shopping list) stand out as everyday life genres you already know very well. Yet, in your professional life you may very well need to write formal business letters and reports of different kinds. Each of these genres has its own characteristics and conventions that make it a genre and, as engineers, you'll be expected to write them appropriately.

One of our objectives in this book is to provide you with an awareness of the differences in language use that are associated with different contexts: engineers today are expected to be *multiliterate* (i.e. be able to use different registers according to the different communicative situations). As engineers you will soon realize that being literate is not enough and that writing an email to a friend is not the same as writing in a job-related context. Although at this point we are just scratching the surface, you should be aware of what readers will expect your documents to look like. You should then be competent enough and deploy writing skills that allow you to adapt your documents to every writing situation. Being multiliterate in the sense defined above is not usually an easy task because you need to have a good command of:

- *content* knowledge: technical and scientific knowledge that is transferred to you at university,
- *context* knowledge: you should be sensitized about the importance of the scientific community or academic context in which your documents will be read,
- *English language* knowledge: level of proficiency in terms of syntax, grammar, vocabulary, etc. in general-purpose English *and* in technical English,
- *genre* knowledge: knowledge of the different written genres used in the technical professions, and
- *writing process* knowledge: knowledge of the most efficient writing skills and techniques for a writing task.

See how the above categories of knowledge can help you identify some of your knowledge gaps and self-assess your current level of writing competence at this very initial stage. More specifically, try to find out with which categories you would encounter difficulties when writing the documents below:

- ✓ Request for detailed figures of faulty end products
- ✓ Evaluation of a machine breakdown
- ✓ Laboratory report
- ✓ Departmental monthly report
- ✓ Report on a meeting or visit
- ✓ Newspaper article
- ✓ MSc final project or thesis
- ✓ Technical manual
- ✓ Brochure
- ✓ Journal (research) article
- ✓ Email to a business contact
- ✓ Letters of rejection, complaint, etc.

Finally, there are different techniques that can help you improve your writing skills as engineers. For example, it has been demonstrated that reading plays a crucial role in learning a foreign language and, most importantly, that good readers make good writers. Reading is very beneficial, but only if you read voluntarily, extensively and for pleasure. As you can imagine, however, reading is not enough. Apart from reading, you should also write and write because while you are learning to write you are also writing to learn the language and to be an efficient communicator. The more you read and the more you write, the better writers you will become. Last but not least, it can also be very helpful for you to acquire some autonomy to allow you to actively participate in your learning process, for example by monitoring your learning and choosing the tasks that best suit your needs and preferences.

CHAPTER 1

What is technical writing?

1.1 Why is it important to study technical and professional communication?
1.2 Characteristics of good technical writing
1.3 Functions of technical discourse

Reflecting on...

Do you think communication skills are of minor importance in scientific and technical studies?

Do you think a technical student can write as well as a humanities student?

What characteristics do you think distinguish a technical text from a non-technical one?

How can your knowledge of general purpose English help you towards writing technical documents?

1.1   Why is it important to study technical and professional communication?

In a world of rushing and pressure to save time, writing documents seems slow and time-consuming. Why write a letter or a memo if you can make a quick phone call? Why spend time thinking about how to put into words information that can be transmitted spontaneously without the extra effort of heeding syntax and punctuation? This logical reasoning fails, though, when we come to consider the type of documents technical writers need to develop as well as the audience they are addressed to. On many occasions, communication is not just from one emitter to one receiver but rather from one to many, as is the case of memos addressed to company staff, or a report meant to be read by more than one person, for example. In addition, most documents generated in the technical field include information that cannot be easily transmitted unless it is orderly displayed on a document. In other words, oral communication may fall short when we need to transmit the information technical documents require. Hence, writing skills can be considered an important factor in the technical and scientific field because:

1. *In many different types of work, writing constitutes an important part of the everyday workload.* In a company, people write to inform about a project or activity (progress reports), to help managers in decision-making (recommendation reports), to communicate within the organization (memos), to ask questions (inquiry letters) and to contact colleagues, distributors, and mates in the same workplace (email messages). These various tasks reveal that writing is a key activity for many technical professionals.

2. *They facilitate communication with co-workers, clients and supervisors, that is, inside and outside the workplace.* Engineers and scientists' writing skills must be of a high standard in order to effectively communicate with the people with whom they work. It is not enough for them to be technically good, they must be skilful in communicating what they are doing and why it is important. As a last resort, their technical and professional value will very much depend on their capacity to convince others of the importance of their work.

3. *They are necessary for a successful career.* Organizations know the advantages of a well-written document since the way they construct their documents reflects their image. Poorly written documents will reveal not only writers' inefficiency but also organizations' lack of seriousness. Thus, engineers who can communicate their thoughts clearly and efficiently are bound to be promoted to more challenging positions. Additionally, being good at written communication skills (in whatever language) is likely to act as an *added value* that enhances your curriculum vitae and helps you stand out from other applicants in a job selection process.

4. *Writing skills contribute to saving time and money.* Good technical writing saves time and, therefore, money. If you create a document, a report, for example, for your superior, which is clear and easy to understand, no time will be wasted on pondering the meaning. In

addition, the higher the position technical professionals hold, the more time they devote to writing documents.   Hence, if they learn to write well, communication will be more effective, they will be more productive and they will save time for the company.

The following text written by an engineer will make you aware of how important communication skills are for an engineer's career.

Stuff You Don't Learn in Engineering School

Newbie engineers often leave school with technical know-how but without workplace savvy

So you've graduated from engineering school, and you've found a good job. Congratulations! But are you really prepared for life as a working engineer? Do you know what steps to take to advance in your field? Do you know how to stay current and competitive? Can you deal with difficult people, like your boss, or clients, or the public? Are you comfortable speaking in front of a crowd? If these issues concern you, rest assured: you aren't alone. Many if not most young engineers emerge from school with fabulous technical talent but little ability in the "soft" skills or even the realization of how important such skills are. They include making decisions, setting priorities, working in teams, running meetings, and negotiating. I can't blame the engineering schools for not covering this ground; after all, they have their hands full just teaching the latest technology.

But when students eventually hit the workplace, they may find their soft skills woefully undeveloped. Who will teach them then? That's the goal of this series: to acquaint readers with the most important nontechnical skills that every engineer needs to be more effective in the workplace and happier in life.

"Write that!"

Let's start with effective communications, by which I mean writing, speaking, and listening. This triumvirate can be said to be the Achilles' heel of engineers—and with good reason: we didn't have to do much writing or speaking in engineering school, and we may not be called upon to do much on the job. But that doesn't diminish their

importance; without them, you'll never be able to convey to people—your boss, your clients, your family, the public—the merits of your work, ideas, and aspirations.

What things do you write now? Memos, reports, business letters, specs, and, of course, e-mail—a lot of e-mail. Writing those kinds of documents may not require the same level of finesse as, say, a novel, but it still takes some of the same skills.

For starters, use clear, simple, direct language. Don't feel that you must always use technical language, jargon, or acronyms. Highly technical language certainly has its place, but it's often the case that your audience isn't technical—for example, a manager, a salesperson, or an elected official.

A great way to improve your writing is to read more. Yes, read more! If you don't have the time, then make the time. Almost any kind of reading is worthwhile—a good newspaper like The New York Times, a non-fiction book or a novel, The New Yorker, The Economist, whatever interests you. The more you read, the more you will notice and appreciate good writing. And over time, it will rub off, improving your vocabulary, your grammar, and your storytelling ability. Plus, you'll probably enjoy the relaxation that comes with reading.

Another tried-and-true technique is to get a peer to critique your writing. You may feel funny about asking, but it's really not such an outlandish request. What's more, it helps. Say you've been labouring over a report, but you've seen it so many times, you can't tell if what you've written is clear. A friend or colleague will be able to tell you if he or she understands what you're trying to say and perhaps point to sections that need revision.

Whenever I'm stuck on a passage, I imagine my former boss, Bud, reading what I've written, asking me to clarify some point, listening to my response, and then saying, "Write that!"

Speak up

For many people, there's no scarier activity than public speaking. But as an engineer, you must be able to speak to a group of people. I'm not talking about delivering your Nobel Prize address. I'm talking about voicing your ideas at staff meetings, briefing your peers after a business trip, pitching a proposal to clients, presenting a project to the general public. Even if you're the most tongue-tied, introverted engineer around, you can learn to speak better.

Think about driving a car. We aren't born knowing how to do this highly dangerous and complicated task. We learn it, and then over time get better at it by practicing. Ultimately, some of us are better at it than others, and some of us enjoy driving more than others do. But it's a skill that can be acquired. The same goes for public speaking. It doesn't happen by osmosis or wishing for it; you learn it.

The first rule of speaking is to know what you'll be speaking about. If you've been asked to address an unfamiliar topic, then politely decline the offer. If that's not an option, then do as much research as you can beforehand, until you feel reasonably comfortable with the subject matter.

Find out who your audience will be and what they expect to learn. The people who invited you to speak can furnish that information. Your presentation should be pitched accordingly.

PowerPoint has become the speaker's audiovisual aid of choice, so you should at least learn how to use it. If you find it's not for you, investigate other speaking tools.

But whatever tool you choose, make sure you know how to use it before you take the stage. The sight of an engineer fumbling with a PC projector won't reassure the audience and will wreak havoc with your confidence.

Never, ever read your presentation. At the least, you'll bore your audience. And if you've practiced sufficiently, you shouldn't need a prepared script.

It's crucial to stick to the allotted time, if only to allow for a question-and-answer period. You can't assume a clock will be available, so always bring a watch. You can even ask someone in the audience to signal you when you're approaching the time limit.

Don't wait for an invitation to speak; seize opportunities, and practice, practice, practice! Offer to introduce speakers at meetings, give briefings on your projects at staff meetings, and join the speakers' bureau of your organization. For added practice, consider joining the nonprofit group Toastmasters International (http://www.toastmaster.org), which has chapters throughout the world where members meet regularly and practice speaking.

The more you speak, the more comfortable you'll be, and the better you'll get. It's that simple, really.

I'm listening

For many of us, listening is the most difficult aspect of communication. It requires paying attention not just to the facts being presented but to the speaker's feelings and intent, as conveyed through his or her facial expressions, body language, and tone of voice.

I admit, I find this a tough skill to master. While another person is speaking, my brain is often busy planning what to say next or distracted by other thoughts. Becoming a good listener may mean breaking some lifelong habits in how you converse.

The keys to improving your listening are to reduce or eliminate any distractions, to make eye contact with the speaker so he or she sees you're paying attention, and to respond appropriately—nod your head in agreement, say "I see your point," take notes. Try not to interrupt the other person, or complete his or her sentences.

If you don't understand what's been said, don't hesitate to say so. ("I'm not clear on that. Could you explain it again?") Try not to think about your answer until the speaker is finished; you can always pause before responding. ("Let me think about that before I answer.")

Not all listening is done face to face, of course. You also need to listen while on the phone and even when reading e-mail. The point remains the same, though: pay attention and respect the speaker.

Can you be an engineer and not be able to write, speak, or listen effectively? Sure—but not a successful one. Communication is one of the most basic and important of the soft skills, and with the practice and patience, you too can become a good writer, speaker, and listener—even if you didn't learn that stuff in engineering school.

Reprinted with permission from Selinger, C. Stuff you don't learn in engineering school. *IEEE Spectrum*, September 2003, 49-52. © 2003 IEEE.

COLLABORATIVE TASK

1-1 After reading the text above carefully, discuss it in pairs and answer the following questions:

a) According to the text, many recently graduated engineers lack some training in non-technical skills. Why is this so? What are these skills? Do you agree with the writer in that they are as necessary as technical skills?

b) The text focuses on three communication skills—writing, speaking, and listening. For each of the skills below, enumerate the reasons why they are necessary and then the ways in which they can be practised or improved, as suggested by the writer.

    WRITING: necessary to:
                    can be improved by :

    READING   necessary to:
                    can be improved by :

    SPEAKING: necessary to:
                    can be improved by :

    LISTENING: necessary to:
                    can be improved by:

c) What are the most important difficulties you find when writing in English? Do you apply any of the techniques recommended by the writer to overcome them? If not, which ones do you apply or can you think of?

d) Do you agree with the writer's final conclusion? Why? Why not?

1.2 Characteristics of technical writing

What is technical writing? To this question there is no simple answer as technical writing entails many aspects: it is writing that includes many different types of documents, that is aimed at different types of audiences and that has many different purposes. Technical writing is usually related to business, industry, administration and technical professions and responds to a need for some sort of action such as documenting, solving problems, helping in decision-making, reporting, performing tasks, instructing, asking and answering questions, etc. Hence, technical writing can be said to arise from a need and to serve as an *action-link* between the writer and the reader (see Figure 1.1). For reasons of brevity, we could say that technical writing is job-related; this includes both academics, for whom writing research articles is part of their professional duty, and engineers working in the industry. The information transmitted usually combines concepts and ideas with numerical figures and visual aids in order to help the reader readily understand the message, so it stands to reason that the language used to convey all this information will also be utilitarian and concise. As opposed to fiction, technical communication does not seek to entertain or amuse the audience but to inform or document them as objectively and efficiently as possible. Technical communicators also aim to be credible, so they contrast their information, present data accurately and provide support to the facts while complying with the conventions and norms of their scientific or technical community.

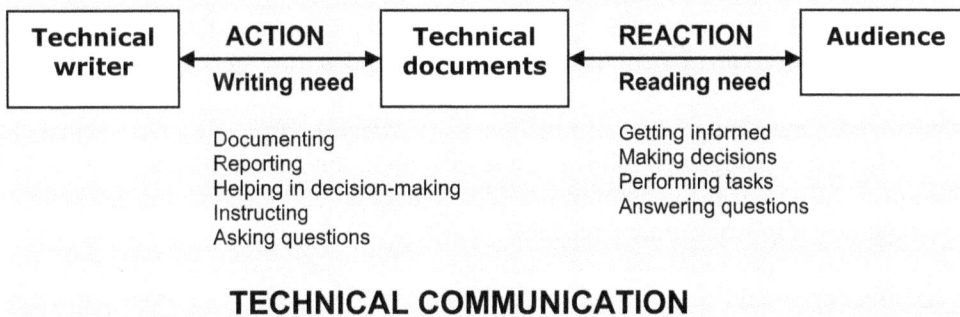

| Technical writer | ACTION
Writing need | Technical documents | REACTION
Reading need | Audience |

Documenting
Reporting
Helping in decision-making
Instructing
Asking questions

Getting informed
Making decisions
Performing tasks
Answering questions

TECHNICAL COMMUNICATION

Fig. 1.1

The general characteristics that portray technical writing are going to be spelled out below, but first there are some defining aspects that should be considered: *content*, *lexis*, *tone*, and *purpose*. Technical writing is mainly defined by its content. The *content* being conveyed is related to scientific and technological matters. Technical discourse deals with objects, instruments, devices, mechanisms, machinery, experiments, properties, systems and processes of different kinds. In order to study and analyze all these technological matters, scientists and engineers often resort to factual and numerical information which usually go hand in hand. The *lexis* is also another distinguishing characteristic, arising from the content, which introduces the specific words and expressions proper of the technical register. As mentioned above, the language is utilitarian, to-the-point, and focused on the content rather than on the author's feelings towards the topic. This implies that the *tone* is usually objective. Finally, with regard to *purpose*, the main and overtly manifest intention of technical writers is to transmit information. This informative purpose, which gives priority to clarity and accuracy, is mainly triggered by the professional or academic setting where technical communication usually takes place.

Keeping in mind the defining aspects mentioned above, we will now look at the main characteristics of technical writing.

Accurate. This is related to the quality of content; that is, information must be true and validated. An error-free document gives credibility and earns the reader's trust. Technical documents often help in decision-making, so inaccuracies may cost money and be a waste of time. It is important, therefore, that you ask for peer review in order to detect any erroneous information.

Thorough. This characteristic refers to the quantity of information to be included, which implies that technical writing must be complete and that important information shouldn't be left out. Likewise, the amount of detail to be included for full comprehension may vary depending on the level of expertise of the intended audience. Writers can also use visual aids to complement the contents of their documents.

Relevant. The content must be closely related to the topic as superfluous, redundant and ornamental information can distract readers from the intended message.

Well-organized. Organizing your information in an orderly, logical way is crucial. Ideas should be arranged in paragraphs following a logical order and in accordance with the line of your argument. Inconsistent jumps from one idea to another and gaps in your reasoning will create confusion and mistrust among readers.

Clear. Information should be presented not only in a well-structured and orderly manner but also in a simple and readily comprehensible way. Therefore, the use of obscure language and verbose style should be avoided as it doesn't contribute to clarity. On the other hand, visual aids are a good means to enhance comprehension.

Concise. *Concise* means expressing your ideas in as few words as possible; technical writers should be succinct and to the point in order not to bore their audience. Remember that *brevity is the soul of wit*, according to the English saying.

Correct. The document should be well-written in terms of spelling, grammar, syntax, and punctuation. A carelessly written document may be associated with poor content and may project a poor image of the writer. Spell checkers, grammars and dictionaries are good tools for helping you write a flawless text.

1-2 Read the following sample texts. You will see that one is clearly technical and the other is not. Bearing in mind the characteristics of good technical writing, justify the reasons for your choice. In pairs, provide as much evidence including examples from the texts as possible.

The Information Age

The advertisers assure us that we are about to plunge into the information age. Exactly what this is I'm not sure, but they say that the era to come will be much better than the Ice Age. Information will be good for us, they add—which is fortuitous since we don't seem to have a lot of choice in the matter. Like it or not we are about to be inundated by streams of information-carrying bits. Information access will be available everywhere —in our homes and offices, through public terminals, and even in our cars and on airplanes. Everyone —excepting those unfortunate individuals and nations who lack the necessary clout in computer IQ— will have the right to bathe daily in tubs of bits, and afterward to pull the plug and watch the excess information drain away. Our time will be spent either accessing information or pretending to access information —the latter activity being recommended to preserve the illusion of the savvy information-age citizen. We shall have to be diligent about this daily hunt for information, since it, rather than material goods, will be basic commodity of the land. Our position in society will be determined solely by our market position in bits.
I'm a little nervous about the coming of this information age for several reasons. First, no one has told me what I'm supposed to do with all those bits. Moreover, I secretly wonder who will be generating all those bits that the rest of us are forced to absorb.
I barely have learned how to take care of money. What happens if we go on the bit standard? Will I be able to find ways to invest my hard-earned bits? Or can I hide bits in my mattress, against the rainy day when I find myself bitless? How will I ensure that I have enough bits to carry myself into retirement, when I no longer have the strength and ability to conduct the arduous, daily accessing of information which will be the sine qua non of the coming age? Will bits depreciate because of bit inflation, or like fine wines will they improve if left in a dank basement? (Perhaps no packet should be opened before its time.) (…)
After Claude Shannon and others conceived the principles of information theory in the late 1940s, a number of studies were conducted to determine the channel capacity of a human being. These studies were fairly consistent and depressing. No one does them anymore. It seems that human beings are only good for about 150 bits per second of input/output. That is all the information we can take in or put out. I ask you, is this creature equipped for the onslaught of the information age?

Reprinted with permission from Lucky, R. W. (1993). *Lucky strikes… again.* New York. IEEE Press, 223-224. © 1993 IEEE.

The CD Digital Audio System

Compact disk (CD) digital audio has undoubtedly introduced a new concept in high-fidelity music recording and reproduction. In this novel audio storage approach, the reproduced sound is virtually unaffected by dust particles, scratches, and fingerprints on the disk.

This is partly due to: 1) digital storage techniques that offer a very high dynamic range, and 2) because the optical recording and reproduction method with a laser beam can access any recording material with pinpoint precision. However, the powerful error control coding scheme that is applied in CD systems can take most of the credit. The high density of data recorded on a CD translates a minimal imperfection on the recording surface of approximately two millimetres into an error burst of 2400 bits in length.

A simple technique that converts long error bursts into random error, or at least into shorter bursts, is called *interleaving*. CD error control coding scheme includes a product code consisting of two RS codes (see Table 4) and two interleaving processes, hence the name of Cross Interleaved Reed-Solomon Code (CIRC).

CIRC permits highly-efficient detection and correction for bursts as well as random errors. For errors that cannot be corrected by CIRC but still can be detected, a technique known as *interpolation* is applied. If two reliable neighbour samples (a group of words or bytes) around an erroneous sample are available, the CD decoder uses these reliable samples to make an estimation (linear interpolation) for the erroneous sample. Listening tests indicate that the degradation introduced to CD systems as a result of interpolation is inaudible. In case there are no reliable neighbour samples, the CD decoder erases the sample entirely. This is known as *muting*. Experimental data demonstrates that as long as muting times does not exceed a few milliseconds, and if it is incidental, muting of the audio signal is inaudible.

1.3 Functions of technical discourse

In the above description of technical communication, we mentioned that this sort of job-related discourse usually arises from a need and serves as an ***action-link*** between the writer and the reader. The purpose and content of technical discourse sharply contrasts with the purpose and content of literary discourse, for example. In fiction or poetry, the writer's purpose may be mainly to entertain, express or create emotions. Conversely, technical discourse usually intends to *initiate* some kind of *action*. As a result, it is commonsensical to find out that content, that is, the kind of information and how this information is organized in technical writing, is greatly dependent on the objectives of technical writing. If we wish to gain deeper insight into the content of technical communication, we will have to analyze technical discourse in more detail.

What are the most common actions or objectives accomplished by technical writing? Technical documents are usually written to carry out actions. For example, technical documents may be written to:

- explain an experiment
- present a hypothesis or theory
- report on past or current research
- recommend
- provide information about a device
- provide information about a procedure
- instruct

  ...

A relevant characteristic of technical discourse is that if we analyze not isolated words but larger chunks, we will soon realize that there is a remarkable number of discourse units (a paragraph, a text, etc.) performing certain functions recurrent in technical communication. Let us now try to figure out some of these functions with the help of common sense and prior knowledge.

Read this text without worrying about unknown words or phrases. As you read it, step back from the text and look at it in a somewhat abstract way, trying to answer: What is this chunk doing? What is the writer doing here?

The dispersal of volcanic aerosols has a drastic effect on the Earth's atmosphere. Following an eruption, large amounts of sulphur dioxide (SO2), hydrochloric acid (HCL) and ash are spewed into the Earth's stratosphere. Hydrochloric acid, in most cases, condenses with water vapor and is rained out of the volcanic cloud formation. Sulphur dioxide from the cloud is transformed into sulphuric acid (H2SO4). The sulphuric acid quickly condenses, producing aerosol particles which linger in the atmosphere for long periods of time. The interaction of chemicals on the surface of aerosols, known as heterogeneous chemistry, and the tendency of aerosols to increase levels of chlorine which can react with nitrogen in the stratosphere, is a prime contributor to stratospheric ozone destruction.

Three types of aerosols significantly affect the Earth's climate. The first is the volcanic aerosol layer which forms in the stratosphere after major volcanic eruptions like Mt. Pinatubo. The dominant aerosol layer is actually formed by sulfur dioxide gas which is converted to droplets of sulfuric acid in the stratosphere over the course of a week to several months

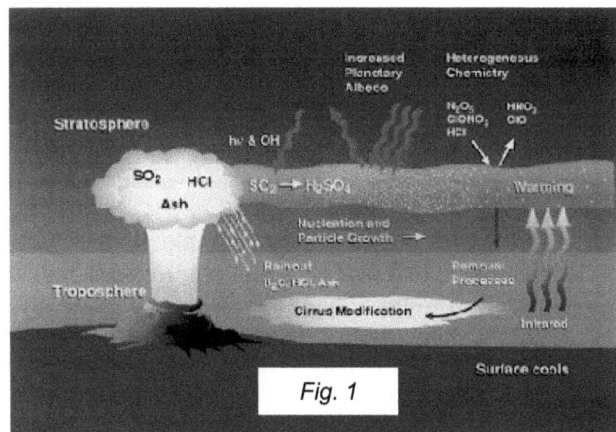

Fig. 1

after the eruption (Fig.1). Winds in the stratosphere spread the aerosols until they practically cover the globe. Once formed, these aerosols stay in the stratosphere for about two years. They reflect sunlight, reducing the amount of energy reaching the lower atmosphere and the Earth's surface, cooling them. The relative coolness of 1993 is thought to have been a response to the stratospheric aerosol layer that was produced by the Mt. Pinatubo eruption. In 1995, though several years had passed since the Mt. Pinatubo eruption, remnants of the layer remained in the atmosphere. Data from satellites such as the NASA Langley Stratospheric Aerosol and Gas Experiment II (SAGE II) have enabled scientists to better understand the effects of volcanic aerosols on our atmosphere. (…)

Source: NASA Langley Research Center. (1996). Atmospheric aerosols: What are they, and why are they so important? [WWW page]. URL http://oea.larc.nasa.gov/PAIS/Aerosols.html) Reprinted with permission of NASA.

You will soon reach the conclusion that the function of this text is to *describe*. In technical discourse description can be presented by means of visual aids or in written form. Description typically falls under three types, namely *physical description*, *function description*, and *process description*. *Physical description* gives the physical characteristics of an object, such as dimension, shape, weight, material, or volume, together with the spatial relations of the parts of the object to one another or to the whole. *Function description* essentially informs readers of the use and purpose of a device. Finally, *process description* deals with processes and procedures and details a series of dependent steps leading to a goal.

Which sort of description can you identify in the text above?

If you look at the text more closely, you'll probably see that the main or primary function is to describe but that in order to expand this description, the writer is also *defining,* and *classifying*. We find a *definition* (*The interaction of chemicals on the surface of aerosols, known as heterogeneous chemistry* …) and an incomplete *classification* of aerosols (*Three types of aerosols significantly affect the Earth's climate. The first is the volcanic aerosol layer which*…). These three functions combine frequently in technical discourse and can be used to extend, support or elaborate on a description.

Finally, another very recurrent function of technical discourse is *instruction*. Instructions can be given in a list or in paragraph form and they can be found in operator's and user's manuals, reports or guidelines for work practices. Here is an example of an excerpt whose function is to instruct the reader.

Place the prepared test sample in the wire basket and immerse it in the water described in 5 at a constant temperature between 15 and 25°C with a cover of at least 50mm of water above the top of the basket.

Immediately after immersion, remove the entrapped air from the sample by lifting the basket containing it 25 mm above the base of the tank and allowing it to drop 25 times at about once per second. The basket and aggregate should remain completely immersed during this operation and for a period of 24 hours. If for special purposes other immersion periods are used, deviating more than 4 hours from the above requirement, this must be stated in the report.

Source: Rilem. (1994). *Technical recommendation for the testing and use of construction materials.* International Union of Testing and Research Laboratories for Materials and Construction. London: E & FN Spon, 46. Reprinted with permission of the publisher.

In a word, even though other functions exist, those of *describing*, *defining*, *classifying* and *instructing* are by far the most frequent in technical discourse. These functions vary in terms of length and degree of detail and do not necessarily coincide with paragraph division, as you have seen; in fact they may overlap or appear in combination quite frequently.

COLLABORATIVE TASKS

1-3 Read the following samples and identify the different functions. Then compare your answers to those of your partner.

Sample 1
Aerosol measurements can also be used as tracers to study how the Earth's atmosphere moves. Because aerosols change their characteristics very slowly, they make much better tracers for atmospheric motions than a chemical species that may vary its concentration through chemical reactions. Aerosols have been used to study the dynamics of the polar regions, stratospheric transport from low to high latitudes, and the exchange of air between the troposphere and stratosphere.

Source: NASA Langley Research Center. (1996). Atmospheric aerosols: What are they, and why are they so important? [WWW page]. URL http://oea.larc.nasa.gov/PAIS/Aerosols.html). Reprinted with permission of NASA.

Sample 2
The steel sleeper is an industrial product of simple construction. It consists of a rolling type profile in the form of ∩. Its ends are forged to provide anchoring in the ballast, so as to ensure transverse track stability. The rail is mounted on the steel sleeper by rail spikes (crampons) fixed by rail spike bolts in holes drilled onto the sleeper top. Elastic fastenings may also be used. Steel sleepers are made from low carbon steel of an ultimate tensile strength of $40\div50kg/mm^2$. Generally sophisticated steels have not been used and therefore the yield strength is near 50% of ultimate strength. The chemical composition is usually 0.15 C, 0.45%Mn, $0.01 \div 0.35$ % Si, $0 \div 0.35$% Cu.

Source: Profidillis V.A. (1995). *Railway engineering.* Avebury Technical: Ashgate Publishing Limited, 117. Reprinted with permission of the publisher.

Sample 3

The ingredients for the Portland cement concrete are Portland cement, a fine aggregate (sand), a coarse aggregate (gravel), and water. In order to produce Portland cement the aggregate particles are made to act as a filler material to reduce the overall cost of the concrete product because they are cheap, whereas cement is relatively expensive. To achieve the optimum strength and workability of a concrete mixture, first the ingredients are added in the correct proportions. Dense packing of the aggregate and good interfacial contact are achieved by having particles of two different sizes; the fine particles of sand fill the void spaces between the gravel particles. Ordinarily, these aggregates comprise between 60 and 80% of the total volume. Subsequently, it is necessary to check that the amount of cement-water paste is sufficient to coat all the sand and gravel particles, otherwise the cementitious bond will be incomplete. Finally, all the constituents are thoroughly mixed with water. Complete bonding between cement and the aggregate particles is contingent upon the addition of the correct quantity of water. Too little water leads to incomplete bonding, and too much results in excessive porosity. In either case, the final strength is less than the optimum.

Source: Callister W. D. (1994). *Materials science and engineering. An introduction*. New York: John Wiley & Sons, Inc., 519. Copyright © 1994 John Wiley & Sons, Inc. Reprinted with permission of the publisher.

Sample 4

The take-off of an aeroplane begins with a very short period during which the tail is raised from the ground. This is achieved by the moment of the propeller thrust T with respect to the point where the landing gear touches the ground. The moment Tt acting clockwise will outweigh the opposite moment Wl due to the gravity at a comparatively low engine speed. Then with the tail raised, the take-off begins; its purpose is to accelerate the airplane to a velocity at which climbing is possible.

Source: Von Mises, R. (1945). *Theory of flight*. New York: Dover Publications, Inc., 469. Copyright 1945. Reprinted with permission of The McGraw-Hill Companies.

Sample 5

Concrete laid in the open air or direct sun should be covered with burlap, roofing felt or building paper during the curing period. This protective covering should be removed before the concrete is wet down. Never attempt a big concrete job on an extremely hot day. Concrete will set up extremely fast in direct sunshine. It is always better to wait until mid-afternoon even if this means working late in the evening.

Source: ACE. (2006). Pouring concrete. [WWW page]. URL http://www.acehardware.com/sm-pouring-concrete--bg-1283398.html

Sample 6

Wastewater engineering is that branch of environmental engineering in which the basic principles of science and engineering are applied to the problems of water pollution control. The ultimate goal—wastewater management— is the protection of the environment in a manner commensurate with the public health, economic, social, and political concerns.

Source: Tchobanoglobus, G. & F.L. Burton (1991). *Wastewater engineering, treatment, disposal and use.* (3rd ed.). New York: McGraw Hill, 1. Copyright 1991. Reprinted with permission of The McGraw-Hill Companies.

Sample 7

Steels are iron-carbon alloys that may contain appreciable concentrations of other alloying elements; there are thousands of alloys that have different compositions and/or heat treatments. The mechanical properties are sensitive to the content of carbon, which is usually less than 1.0 wt%. Some of the more common steels are classified according to carbon concentration, namely, into low-medium-, and high-carbon types. Subclasses also exist within each group according to the concentration of other alloying elements. *Plain carbon steels* contain only residual concentrations of impurities other than carbon and a little manganese. For *alloy steels*, more alloying elements are intentionally added in specific concentrations.

Source:.Callister W. D. (1994). *Materials science and engineering. An introduction*. New York: John Wiley & Sons, Inc., 353. Copyright © 1994 John Wiley & Sons, Inc. Reprinted with permission of the publisher.

Sample 8

The Command Module 4m (13ft) in diameter and 13.6 m (12 ft) high, contains a crew compartment for three astronauts, a docking tunnel to the top of the cone-shaped module, and a hatch that can be opened from the inside after docking with the Multiple Docking Adapter. Twelve latches at the outside of the tunnel end will attach the Command Module firmly to the port of the MDA before crew transfer can begin (Fig. 103).

Source: Belew L. F. & E. Stuhlinger (1973). *Skylab. A Guidebook*. George C. Marshall Space Flight Center, National Aeronautical Space Administration, US Government Printing Office, 81.

1-4 Write a short description of a VDM (Visual Display Monitor) giving as much information as possible. Try to combine the three different types of description and, if possible, include an illustration to help readers better understand your text.

1-5 Carefully read the table below and choose one of the engineering degrees you prefer or that is most closely related to your field of study and interest. Then write a short paragraph including a definition and a classification of the items involved.

Degree	Fields/Specialization
Civil Engineering	Transport
	Urbanism
	Hydraulic engineering
	Maritime engineering
	Geotechnical engineering
	Analysis and design of structures
	Technology and construction of structures
	Sanitary and environmental engineering
Telecommunications Engineering	Telematics
	Communications
General Engineering	Industrial engineering
	Materials engineering
	Chemical engineering
	Mechanical engineering
	Electronic and electrical engineering
	Environmental engineering
	Product and system design
	Logistics
	Bioengineering
Computer science and engineering	Computers
	Networks
	Operating systems
	Software engineering and information systems
	Industrial information technology

1-6 Study the table below showing characteristics and typical applications of major materials in engineering. Choose one type of material, or a subtype, that you know is commonly used in your speciality or field of study. Then with the help of the information in the table write a short paragraph using some of the functions seen in this chapter:

Material Type	Major characteristics	Typical applications
Metals (aluminium alloys, steels, iron, titanium alloys, etc)	High electrical and thermal conductivity, corrosion-resistant, strong, ductile, mechanical properties.	Automobile axles, valves, gears, pistons, engine blocks, space vehicles, airplane wings, missiles, aircraft, steam boilers, heat-treating furnaces, gas turbines, railroad rails and wheels, bridges, bus bodies, pipe fittings, I-beams, etc.
Ceramics (clay minerals, cement, abrasives, glass, refractories advanced ceramics).	Optical transparency, easy to fabricate, strong, durable; resistant to thermal shock and high thermal conductivity.	Oven ware, laboratory ware, table ware, optical lenses, heat engine applications. Used in the electronic, computer, communication and aerospace industries.
Polymers (plastic and rubber)	Rigid and brittle, flexible and elastic, abrasion resistant, high tensile strength, high modulus of elasticity.	Coatings, adhesives, foams (thermal insulation and packaging), films, electrical moldings, laminates, curtains, tubing, raincoats, bearings, gears, lenses, refrigerator linings, toys, pipe, electrical wire insulation, fibreglass boats, sinks.
Composites (cermet, concrete, fiber-reinforced, hybrid, laminar	Property combination to improve and reinforce stiffness, toughness, and ambient/high temperature strength, etc.	Concrete as a paving material or as a structural building material (in construction, highway and railway bridges), aircraft, automotive and marine bodies, industrial floorings, storage containers, new engine applications, sporting goods, lightweight land, water, and air transport structural components, orthopaedic components, the modern ski.
Semiconductors	Electrical properties (ability to amplify an electrical signal and act as switching devices); low power requirements.	Electronic devices like the small silicon chip, transistors, diodes, capacitors. Calculators, watches, industrial production and control, microelectronic circuitry, miniaturized circuitry, computers.

1-7 Drawing on your technical knowledge, write a short paragraph about ONE of the topics below. Your paragraph should include a classification and either a definition or a description of the different items involved.

TOPICS: Computers, programming languages, antennas, mechanical engines, bridges and steels.

1-8 Choose one of the paragraphs written in the previous tasks. Then read it aloud to the rest of the class so that they can identify the different functions you have used.

1-9 Write a short paragraph summarizing the text *Stuff you don't learn in engineering school* on page 15.

PART II

THE WRITING PROCESS

Introduction to the writing process

Try to remember the last time you had to write a more or less formal (academic or professional) document. With the help of the questions below, reflect on your usual writing habits and their usefulness.

❑ Did you do anything before beginning to write (for example, mentally scan the main ideas you wanted to transmit and/or jot them down, look for information, schedule your work in terms of time, outline before or after your first draft)?
❑ What did you do when writing (simply sit in front of the computer, create a new document and begin writing your final version, write several drafts)?
❑ What did you do once you had completed your first version (allow for thorough revision, quickly scan for any mistakes, print it and hand it in)?

Beginning to write may be a hard task for most people as ideas come mixed up in a disorderly manner. In trying to get started, many different aspects come into mind: content, style, grammar, etc. and it may be difficult to cope with them all at the same time: In order to seek guidance and to acquire confidence, the writer may find it useful to resort to some kind of systematic and integrative approach which takes into consideration the most important aspects of writing.

The integrative approach adopted in this book draws on knowledge of different approaches to writing (see Figure 1). On the one hand, it takes into account the linguistic knowledge about texts, namely, grammar and text structure. Mastering syntax, an appropriate use of vocabulary and cohesive devices as well as patterns of information organization become essential to produce well-written texts. This is known as *product approach*. On the other hand, the integrative approach also pays attention to the writing skills or stages involved in writing. Novice writers should be made aware of writing as a process consisting of different stages (planning, drafting, revising, etc.) when creating a text (*process approach*). Finally, this approach also heeds the social context, mainly the purpose and audience the document is addressed to, as well as the writing conventions established by the technical and scientific community (*genre approach*).

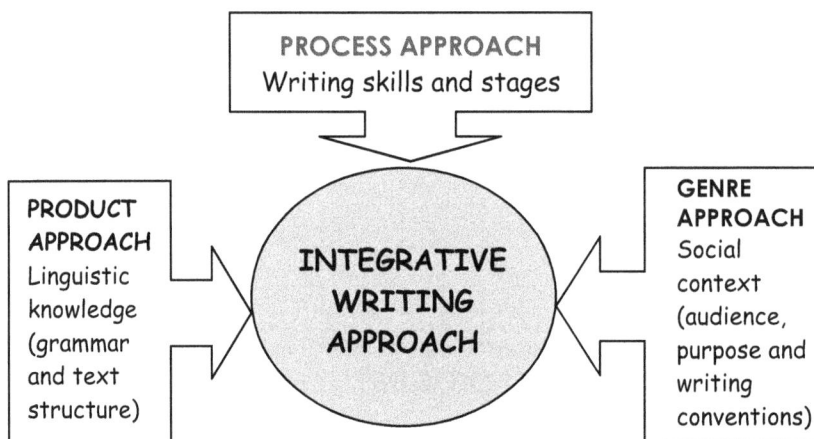

```
        ┌─────────────────────────┐
        │    PROCESS APPROACH     │
        │  Writing skills and stages │
        └─────────────────────────┘
```

┌──────────────┐   ┌──────────────┐
│ PRODUCT   │   │ GENRE   │
│ APPROACH   │   INTEGRATIVE   │ APPROACH   │
│ Linguistic │   WRITING   │ Social   │
│ knowledge   │   APPROACH   │ context   │
│ (grammar   │   │ (audience, │
│ and text   │   │ purpose and │
│ structure) │   │ writing   │
└──────────────┘   │ conventions)│
                                                         └──────────────┘

Fig. 1

At this point it is useful to clarify that the three approaches mentioned above will be combined into one by subsuming the product and the genre approach under the process approach which, in turn, will serve as the guide to organizing the information in the following chapters. However, the writing approach presented below is by no means intended to be prescriptive. Instead it has been designed to provide guidance allowing enough room for manoeuvre so that writers can adapt these guidelines to their own writing preferences and style. We view writing as a non-linear and recursive process composed of three main stages:

1. *Pre-writing.* Before beginning to write you should invest some time planning what to write and how to transmit the information. In order to do this you should consider (a) *audience and purpose* (who you are writing to and why), (b) *tone and style* (how you transmit the information), (c) *gathering of information* (brainstorming, analysing sources of information, etc.) and (d) *outlining* (organization of information).

2. *Writing*. Once you have gathered and organized the information, you can begin writing a first draft. At this stage, it is important to consider the main parts of the text, paragraph development and coherence as well as genre conventions. As you revise and consider all these aspects, it may be helpful to use representative models as a reference.

3. *Post-writing*. The final stage of the writing process involves (a) *revising content and organization*, (b) *checking for grammatical accuracy* (c) *editing for style and* (d) *proofreading and peer review*. These steps will help you spot any inconsistencies in your document so as to produce a flawless final version.

The three main stages of the writing process together with their corresponding substages are shown in Figure 2 below.

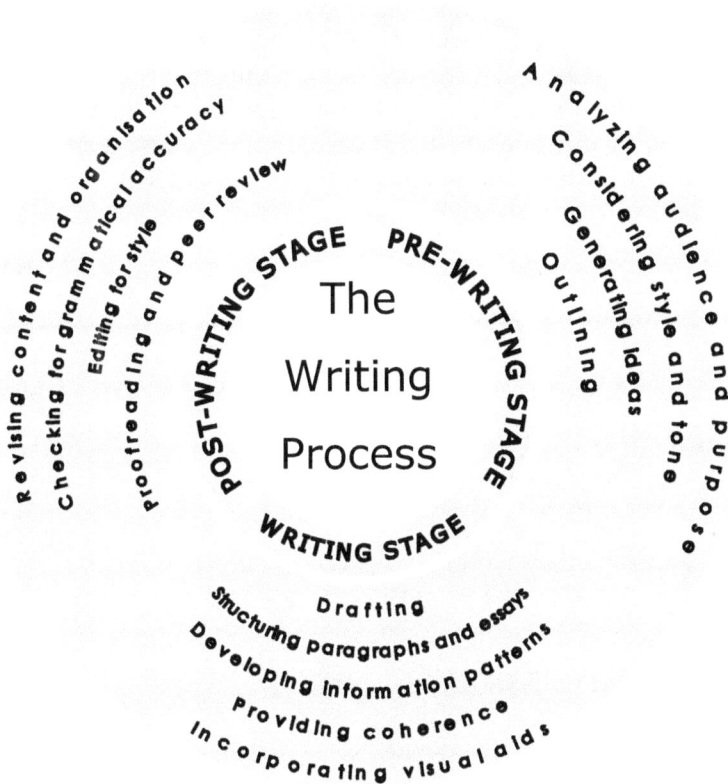

Fig. 2

Some of the benefits that can be obtained from adopting this process approach are outlined below:

- It helps the writer overcome the blank page syndrome and therefore get started.
- It serves the writer as a guide to writing since it suggests possible steps to follow in the writing process.
- It makes the writer aware of contextual considerations such as audience and purpose.
- It promotes awareness of the writing process.
- It accounts for individual variation, that is, it encompasses different learning styles and preferences.

The chapters that follow develop in detail the three main stages of the writing process–pre-writing, writing and post-writing—to help you improve your writing skills. Chapter 2 focuses on the pre-writing stage, in which you must examine your purpose(s), determine your audience, consider the style and tone to adopt, gather data and decide how to organize information. Chapter 3 is based on the writing stage itself. In this chapter you will learn to develop paragraphs, to order information and to provide coherence to your document while drafting your text. Chapter 4 deals with the final stage of the writing process, the post-writing stage. This stage is essential for successful writing as it allows you to polish your document for a perfect final version. For practical purposes, the three stages of the writing process are described in this book in the order described above but remember that this process is dynamic and flexible, and that the different stages often overlap. Therefore, you may go back and forth at your convenience while you draft your document.

CHAPTER **2**

Pre-writing stage

Reflecting on…

By and large, writers are usually recommended to bear in mind 'contextual factors' when it comes to writing a text. Can you guess what these 'contextual factors' might be?

Imagine you have to write a document that describes the latest improvements on a particular product. How may this document differ when aimed at the head of the technical department and the general public? Could you provide a list of different groups of readers an engineer may address his/her documents to?

How is your attitude towards the topic and your relationship with readers reflected in your writing?

Can you think of different ways of organizing ideas before writing a document? Which method do you usually prefer? Why?

2.1 Introduction

Imagine you have just been told to write a short report. What is your usual reaction? Do you sit in front of the computer, create a new document and begin jotting down sentences as ideas come to your mind? If this is what you do, you have a slim chance of writing an appropriate, coherent and effective document. Unless you are an expert writer, you should spend some time on what is known as the *pre-writing stage*.

The pre-writing stage is a very important stage with many aspects to be considered before you actually begin writing. As you will make decisions that will affect and determine the content, approach, or structure of your document, it pays to devote time to answering relevant questions and seriously considering the different alternatives available. In the pre-writing stage the following questions should be addressed:

- WHO am I writing to and WHY?
  Consider audience and purpose.

- HOW should I transmit the information?
  Consider tone and style.

- WHAT ideas should I include in my document?
  Gather information (by generating ideas, analyzing information sources, etc.).

- HOW should I organize and structure this information to best suit the audience's needs and to accomplish my purpose?
  Outline (organize information).

2.2 Analyzing audience

Before writing a document you need to know your audience in order to satisfy its needs. It may be a good idea to develop a profile of the audience you are addressing by answering the following questions:

Who is going to read the document? Here you should analyze the audience's characteristics such as educational and cultural background, position within the company and English competence. In addition, you should take into account whether the text is addressed to a single person or a large group and consider the possibility that a secondary audience might also read your document.

Why does this audience need to read the document? Find out the reasons of the audience's reasons for reading the document. Consider likely actions to be taken after reading (for example making a decision, acquiring new knowledge, performing a task), whether the needs and objectives are job-related or personal and whether they are short-term or long-term. These choices will determine the approach of your document, the information to be included and in what order it should be presented.

How much does the audience know about the topic? Also assess the audience's level of expertise and decide if you need to include introductory conceptual frameworks (conceptual background) and definitions for non-expert audiences, or non-defined technical acronyms, abbreviations and jargon for more specialized audiences.

What is the audience's attitude towards the topic? Analyze the audience's receptivity and attitude towards the topic, the conditions under which the document will be read and the importance given to the topic because these factors will also determine what to say and how to say it.

Types of audience

The answers to the questions above will provide a profile of the main types of audiences in the scientific and technical fields, namely technical experts, technicians, managers and administrators, laypeople, students and mixed audiences. These audience profiles will help you identify your audience in order to write an appropriate document.

Technical experts

This group includes high-tech readers who understand the technical vocabulary, the concepts and the implications related to the field you are writing about. They are readers that:
- read to know about the latest discoveries and contributions to update their technical knowledge.
- understand jargon, acronyms and abbreviations.
- show interest in theory and want details.
- look for graphical information to learn about results.
- read critically and expect well-supported claims.
- are mainly interested in the method and conclusions.

Technicians

These are readers whose job involves skilled practical work with technical equipment and who
- read to find out *how to* perform technical tasks.
- are more interested in practical aspects than in theoretical ones.
- have a good understanding of technical vocabulary.

- may have a limited knowledge of the theory unless they have a higher level.
- may seek some background information to increase understanding and to better accomplish their tasks.
- need and want visual information.
- read thoroughly and follow explanations to the letter.

Executives and administrators

This group is made up of people in a business, institution or organization with administrative or managerial powers. Executives and administrators
- read to make decisions.
- have some familiarity with the technical vocabulary of the field.
- read selectively because they are pressed for time.
- want to find essential information easily.
- are interested in the gist of the document, that is, recommendations and conclusions.
- look for graphics and other visuals.
- want generalizations, not details.
- want information to be explained in plain terms.

Laypeople

Laypeople or the general public are readers that are not acquainted with the subject matter or field of expertise. These readers
- are interested in furthering their knowledge of those technical topics that may have an impact on their lives.
- do not understand abbreviations, acronyms and jargon.
- need background information.
- require simple explanations with descriptions and definitions.
- want graphics to better understand information.
- are more interested in practice than in theory.

Students

This group includes readers that want to learn about a field of study in order to get a degree. In the future, they will become experts in that area, so they
- read for learning.
- need and want theory.
- can understand some technical vocabulary.
- expect numerical and graphical information.
- require clear explanations through definitions and examples.
- are interested in generalizations and details.
- appreciate a friendly and didactic presentation of information.

Mixed

This audience consists of a combination of readers from the aforementioned audiences. Because this group includes readers with different levels of understanding and different purposes for reading, texts addressed to this group should
- include background information.
- define technical terms.
- display a matter-of-fact, businesslike tone.
- be clear and to the point in order to cater for diverse levels of expertise.

These different types of audience profiles are all considered to be *primary readers*, as they usually take some action or make decisions after reading the document. However, you should bear in mind that there may also be *secondary readers* who read for purposes other than the main one and who do not necessarily take any action after reading the document. Imagine the head of a financial department who reads a report to make a decision on the purchase of new shares that may increase company benefits in a near future. This head of department is a primary reader as he/she is authorized to take some sort of action. However, a head of a different department in the same company who also has access to the document but who lacks the authority to take any action would be a secondary reader. Although it is sometimes necessary to make adjustments to cater for the secondary audience, you should mainly concentrate on primary readers.

The following samples illustrate how a text can be accommodated to a given type of audience.

Outbound Channel Protocol

The outbound channels operate in a point-to-multipoint broadcast mode. The most efficient way to utilize these channels is to simply order all outbound packets in a queue of appropriate service discipline. To facilitate channel frequencies and bit-timing synchronization at the MTs, the outbound broadcasts employ continuous transmissions, with filler bits inserted when no data packets are available for transmission. A TDM framing scheme is employed in each outbound channel and is closely related to the framing and slotting scheme in the corresponding inbound channel, which will be introduced in the next subsection. The duration of each frame is of the order of one MT-to-Data Hub propagation delay. Since the data rate is only 2400 bps, each frame is only several hundred bits long. Several consecutive frames are grouped to form a masterframe, and overhead bits which enable the MTs to maintain inbound channel time slot, frame and masterframe synchronization are transmitted once every masterframe. So that an Mt can "switch on the fly" between different data channels without losing frame synchronization, all outbound channels from the same Data Hub employ the same framing format and are synchronized with respect to frame and masterframe boundaries.

This extract is a clear piece of technical writing addressed to an audience of technical experts. The main feature to support this and that stands out at a first reading is the technical vocabulary used. This text is full of jargon and compound nouns that only a person who really knows about the topic is able to understand. Closely related to vocabulary is content, which in this sample is difficult to grasp owing to the technical vocabulary used. In fact, a non-expert will perceive that some sort of process is being described here, yet the actual process and its details can only be understood by someone who is truly involved in the field. As usual, bare acronyms appear; that is, acronyms with no explanation or definition to help the reader decipher them. Also, as in most texts aimed at experts, we find some typical Latin verbs usually associated with formal style.

Maintenance and Repair Guidelines

Maintenance and repair methods
Many different methods of structural maintenance and repair are available. The methods listed below cover the various scenarios experienced during the ship's service life:

Repairs of existing coated areas in segregated ballast tanks

No coating repair:
Coating repairs are not necessary if the coating condition is GOOD or FAIR and the corrosion wastage does not exceed acceptable allowances within the intended service life. Further inspections might be advisable to verify the effectiveness of this coating.(...)

Soft coating repair:
Soft coating repair should only be considered as a temporary maintenance method and is not considered equivalent to the renewal of hard coating. Soft coatings may have to be removed for survey. Annual inspection would be required to verify the soft coating effectiveness and continued structural soundness. (...)

Hard coating repair:
Hard coating in POOR condition should be renewed prior to corrosion wastage exceeding acceptable corrosion allowances.

Anode replacement or new installation:
Existing wasted anodes should be replaced or new anodes can be installed to reduce future corrosion. Anodes only become effective while immersed in water and should only be used as a secondary corrosion protection method.
Anodes should be renewed on the basis of the consumption rate and the remaining anode material.

Source: Tanker Structure Co-operative Forum. (1997). *Guidance Manual for Tanker Structures*. London: Witherby & Co. Ltd., 77-78. Reprinted with permission of the publisher.

If we take into account that technicians read mainly to find out how to perform an action, by just reading the title of this text we immediately sense that the text is aimed at this audience. If we go on reading, our first impression that the text is addressed to technicians is confirmed since its content is based on a series of indirect instructions on whether different types of coating repairs are necessary. The vocabulary used consists mostly of general vocabulary and the few technical terms such as *corrosion wastage* and *wasted anodes* are well-known to technicians working in the field. In addition, for the purpose of clarity, the format of the text includes different headings and subheadings to ease reading comprehension.

Appendix A
Speer Industries, Inc. ©

To:	Mr James Young
From:	John Wayne, Reliance Instrumentation Company
Subject:	Strategic Planning Task Force

This is in reply to your memorandum of October 3. Enclosed is the following information:

A revised statement of the current strategy for each business of Reliance.
A three-year forecast of sales for each business.

As to the next steps, I suggest that we first identify the market segments most likely to achieve the largest growth—say, in the next ten years. And, that we also assess the capabilities of Speer as they might be utilized in entering and/or expanding in these markets.
The Task Force members from each division could be asked to act as a subcommittee to prepare nominations which should include:

The reason why the market segment is expected to grow at that rate.
The capabilities of Speer that could be utilized in these markets.
The capabilities of competitors or likely competitors in these markets.

These nominations could then be reviewed by the entire committee to pick the most likely candidates in light of Speer available resources. In-depth studies of the most likely candidates could then be commissioned under the direction and control of the Task Force.
Assuming some favourable recommendations would be indicated by the in-depth studies, detailed plans of alternatives could then be prepared for final decision by top management.
Plans for the existing business follow: (...)

Source: Abell, D. & J. Mammond. (1979). *Strategic market planning. Problems & analytical approaches.* New Jersey, Englewood Cliffs: Prentice Hall, 480-481. Reprinted with permission of the publisher.

The content of this document clearly reveals that it was written to provide requested information in order to make a decision. In particular, the idea is explicitly stated in the very last sentence. The content as well as the way the information is organized in the text indicate that the audience this document is addressed to is administrators. The main traits indicating this organization are the use of short paragraphs and highlights in the form of lists, which help identify the main points, thus facilitating the reading process. As previously pointed out, administrators are pressed for time and need to find important information easily. The vocabulary used is very appropriate for this audience as the few technical words used are quite common in the business-administration field.

COLLABORATIVE TASKS

2-1 After reading the following texts, decide on the audience to which they are addressed. To justify your answer consider a) the complexity of the content, b) the use of technical vocabulary, c) the use of definitions and exemplification, and d) the use of non-verbal information (numerical and graphical). Then compare your answers with your partner's and discuss.

On Sprites and their Exotic Kin

About 15 years ago, on a clear dark night on the Minnesota prairie, a young scientist testing his auroral imaging camera discovered giant flashes of light illuminating the sky above the distant thunderstorms. Without knowing it at the time, Robert Franz had observed what became known as "red sprites". Researchers wondered how such a spectacular phenomenon, visible to the naked eye and in our immediate surroundings, could have gone unnoticed for so long. Of course, it had not gone unnoticed: Scientists had not paid attention to eyewitness accounts through the years.

The discovery sparked considerable research activity, particularly in the United States, where hot humid air masses sweeping up from the Caribbean and the Pacific power frequent summer thunderstorms. The Rocky Mountains serve as a perfect platform from which scientists during the night can point their sensitive video cameras over the thunderstorms on the plains. Over the past

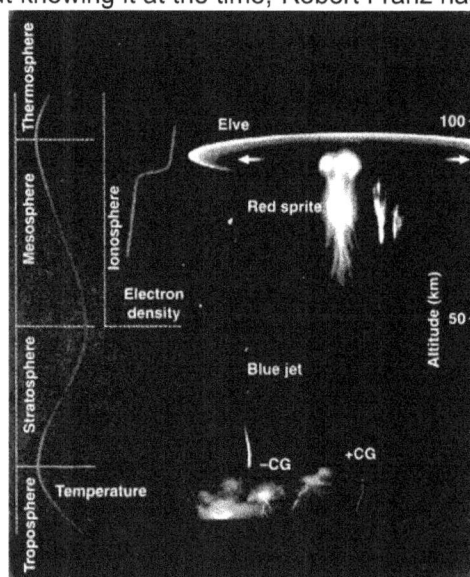

decade, such studies have turned up a surprising collection of optical emissions above thunderstorms, such as "blue jets" and "elves" (see the figure). The question of whether sprites and jets affect the atmosphere in important ways—for instance, through altering greenhouse gas concentrations in the stratosphere and mesosphere or modulating the atmospheric electric circuit—is receiving increased attention.

But what are sprites? They are luminous flashes that last from a few milliseconds to a few hundred milliseconds. The larger sprites may reach from 90 km altitude almost down to the cloud tops, extending more than 40 km horizontally. They are often carrot-shaped and made up of bundles of filaments with diameters of 100 m or less, but can also take other forms, such as a collection of vertical columns.

Excerpted with permission from Neuber, T. (2003). <u>On sprites and their exotic kin</u>. *Science,* 300, 2 May,747.

Sources and Signals

A *source of information* generates a *message*, examples of which include human voice, television picture, teletype data, atmospheric temperature and pressure. In these examples, the message is not electrical in nature, and so a *transducer* is used to convert it into electrical waveform called the *message signal*. The waveform is also referred to as a *baseband signal*; the term "baseband" is used to designate the band of frequencies representing the message signal generated at the source.

The message signal can be of an *analog* or *digital* type. An analog signal is one in which both amplitudes and time vary *continuously* over their respective intervals. A speech signal, a television signal, and a signal representing atmospheric temperature or pressure at some location are examples of analog signals. In a digital signal, on the other hand, both amplitude and time take on *discrete values*. Computer data and telegraph signals are examples of digital signals.

An analog signal can always be converted into digital form by combining three basic operations: *sampling, quantizing,* and *encoding*, as shown in the block diagram of Fig. 1.1 In the sampling operation, only *sample values* of the analog signal at uniformly spaced discrete instants of time are retained. In the quantizing operation, each sample value is approximated by the nearest level in a *finite set of discrete levels*. In the encoding operation, the selected level is represented by a *code word* that consists of a prescribed number of *code elements*. The analog-to-digital conversion process so described is illustrated in Fig. 1.2. Part (*a*) of the figure shows a segment of an analog waveform. Part (*b*) shows the corresponding digital waveform, based on the use of a *binary code.* In this example, symbols 0 and 1 of the binary code are represented by zero and 1 volt, respectively. The code word consists of four *binary digits* (bits), with the last bit assigned the role of a *sign bit* that signifies whether the sample value in question is positive or negative. The remaining three bits are chosen to provide a numerical representation for the absolute value of a sample in accordance with Table 1.1.

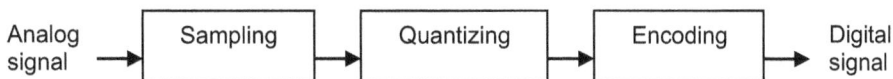

Analog signal → Sampling → Quantizing → Encoding → Digital signal

Fig. 1.1

Source: Haykin, S. (1988). *Digital communication.* New York: John Wiley & Sons, Inc., 2-3. Copyright © 1988 John Wiley & Sons, Inc . Reprinted with permission of the publisher.

2-2 Individually read the following texts and decide whether they were written for a high tech, low-tech or lay audience. Use the table provided to reach a reasoned answer. Then compare your answers with those of your peers to spot any significant differences.

	TEXT 1	TEXT 2	TEXT 3
Content	Technical Sub-technical General	Technical Sub-technical General	Technical Sub-technical General
Language	Acronyms (non-defined) Acronyms (defined) Technical vocabulary (non-defined) Technical vocabulary (defined) Compounds Colloquialisms	Acronyms (non-defined) Acronyms (defined) Technical vocabulary (non-defined) Technical vocabulary (defined) Compounds Colloquialisms	Acronyms (non-defined) Acronyms (defined) Technical vocabulary (non-defined) Technical vocabulary (defined) Compounds Colloquialisms
Format	Visuals Numerical information	Visuals Numerical information	Visuals Numerical information
AUDIENCE TYPE	*HIGH-TECH* *LOW-TECH* *LAY*	*HIGH-TECH* *LOW-TECH* *LAY*	*HIGH-TECH* *LOW-TECH* *LAY*

Discrete Wavelet Analysis

A disadvantage of Fourier analysis is that frequency information can only be extracted for the complete duration of a signal $f(t)$. Since the integral in the Fourier transform equation (4.10) extends over all time, from $-\infty$ to $+\infty$, the information it provides arises from an average over the whole length of the signal. If, at some point in the lifetime of $f(t)$, there is a local oscillation representing a particular feature, this will contribute to the calculated Fourier transform $F(\omega)$, but its location on the time axis will be lost. There is no way of knowing whether the value of $F(\omega)$ at a particular ω derives from frequencies present throughout the life of $f(t)$ or during just one or a few selected periods. This disadvantage is overcome in *wavelet analysis*, which provides an alternative way of breaking a signal down into its constituent parts. The original impetus for wavelets came from the analysis of earthquake records (Goupillaud *et al.*

[107]), but wavelet analysis has found important applications in speech and image processing and is now a significant new tool in signal analysis generally.

$$X(\omega) = \frac{1}{2\omega} \int_{-\infty}^{\infty} x(t)(\cos \omega t - i \sin \omega t)\, d\,t = \frac{1}{2\pi} \int_{-\infty}^{\infty} x(t)\, e^{-i\omega t}\, d\,t$$

Fig. 4.10

Newland, D. E. (1993). An Introduction to random vibration, spectral and wavelet analysis. (3rd ed.) Longman, 295.

They Know Where You Are

The terrorist blast had destroyed the office building. Piles of glass and concrete littered most of a city block, the air was thick with dust, debris still smoldered. The police had no suspects but had already sent out an all-points alert. Then, when troopers pulled a van over for making a couple of risky lane changes, they found a pile of fertilizer sacks and an empty fuel-oil drum in the back. A duffel bag held a change of clothes, a small kit with a new razor and other toiletries, and a 45-caliber pistol. The truck had been stolen, and the driver wasn't talking.

Within an hour, some 1000 km away, an FBI team walked into a Wal-Mart with pictures of the arrested man. One of the cashiers recognized the face. "There were four of them," she said. "One of our regular customers said they were friends visiting from out of town—but that guy's a loner. He lives out on County 15."

This may sound like the start of a mediocre TV drama. But given recent events—and coming technical advances—it just might be a scenario pulled from tomorrow's news. Here's the rest of the story: a bit of microcircuitry called a radio frequency identification (RF-ID) tag was embedded within the package of razor blade cartridges in the suspect's toiletry kit. The manufacturer inserted it into that package, and all others of its kind, to let retailers track inventory cheaply and conveniently. But because the tag carried a unique identifying code, the FBI could scan it, check it against a database, and then track down the store where the razor was purchased.

The future, in this case, is already here. This past January, the Gillette Co. (Boston) announced that it would purchase up to half a billion RF-ID tags to put on its Mach3 and Venus razors and razor blade packages. The tags, which contain chips that respond to an RF field from a scanner, are now being used in a test by Wal-Mart Stores Inc., by the UK-based grocery chain Tesco PLC, and most recently by Metro AG, Germany's largest retailer, to determine whether the technology can streamline inventory management and save retailers billions of dollars a year in supply chain costs.

It's not much of a stretch to imagine this relatively benign way of tracking goods being put to other, more dramatic uses. Indeed, RF-ID tags are only one example of a coming wave of wireless communications technologies that will be everywhere in the next year or two, merging location and time-related information.

The most spectacular of these will be large-scale systems that piggyback on cellular networks to locate any cellphone on the network—in other words, in the near future, your whereabouts won't be a secret if you are carrying your cellphone. Other plans revolve around smaller-scale technology that will, for example, let you buy an item merely by pointing at it with your cellphone.

The commercialization of these technologies promises to make your life safer, easier, and more enjoyable by providing instant, personalized information. It might even save your life by helping rescue officials find you in an emergency, no matter where your are.

But these benefits will almost certainly cost you some privacy. In the case of the tagged razor blades, the loss will be small and incremental; with cellphone tracking, it could be substantial and potentially intrusive. So, in coming months, expect some clashes as watchdog groups, business, and governments try to find common ground and deliver the benefits of location tracking with the least possible intrusion. (…)

Reprinted with permission from Warrior, J., E. McHenry & K. McGee. They know where you are. *IEEE Spectrum*, July 2003 (7), 20. © 2003 IEEE.

Trench Excavation

The amount of excavation required on site for trenches and pits is again determined by means of sight rails and boning rods. The positions of the trenches for the strip foundations of a building are established by means of profiles. Having established the positions of the corners of the building,

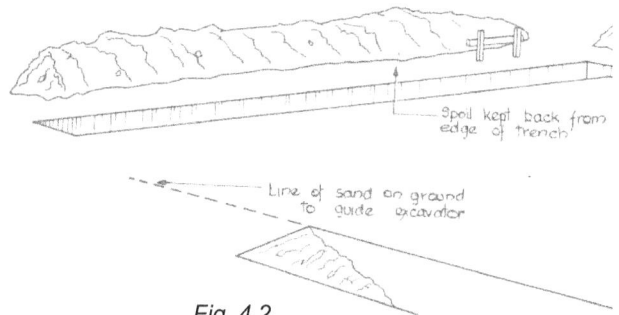

Fig. 4.2

the profiles (horizontal boards nailed to two stout posts driven into the ground) are set up clear of the trenches, so as not to obstruct the excavation work (fig. 4.2). The positions of the trench and walls are marked on the top of the board by nails or saw cuts, so that lines can be strung from profile to profile indicating the exact run and width of trench or wall. If the tops of all the profile boards are kept at the same level, then the profile can also be used as a sight rail (fig. 4.3).

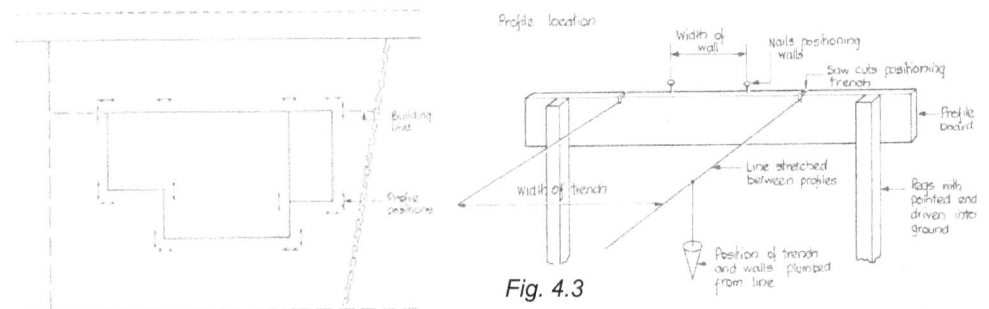

Fig. 4.3

Foundation-trench excavation is usually carried out by means of a back-actor excavator or by hand. As most of the excavated material will be used to backfill the trench on completion of the substructure works, it is deposited alongside the trench, but not so close to the edge as to cause that edge to give way under the additional loading. The mechanical equipment will excavate to the full depth at one pass, whereas with hand excavation several passes will be required, especially in excavations exceeding 1.5 m deep, where the labourer will be unable to "throw out" the excavated material with ease.

Reprinted from Construction technology, (Vol. 2) by Grundy, J.T. (1979), Chapter 4, 26-27, © 1979, with permission from Elsevier.

2-3 In groups of two, write two short paragraphs on a technical topic (a concept, object, method, process, device, etc.) you know well. One should be addressed to an expert in the topic and the other to a person who knows little about it. Also indicate the writing criteria you followed to accommodate each text to its corresponding audience.

CRITICAL THINKING

2-4 With respect to audience, what difficulties may you encounter?
- For example, of the two paragraphs you have written in task 2-3, discuss and decide with a peer which paragraph you found most difficult to write and say why.
- How may writing to people from a different culture affect the comprehension of your text?

INDIVIDUAL TASKS

2-5 You have to write a technical description for a group of students who have not had much training on the subject. Decide on a topic of your field of studies and work on two aspects that should be taken into account before beginning to write:
- *Content*. What are the main ideas and subideas you intend to include?
- *Format*. What are the main titles and subtitles of your text? Do you intend to include any numerical or visual information? Is there anything that should be highlighted?

You may draw an outline with all this information, including the main ideas and subideas under the corresponding headings.

2-6 Imagine you are planning to write the technical description in task 2-5 on the subject chosen. Select the subject-related technical terms you foresee students will not understand, and write a definition for each.

2-7 Write a trip memo for a multiple audience in which you explain the main ideas presented in a conference about the latest discoveries of a technical subject related to your field.

2-8 Imagine a situation where you are working in an audit-consulting company and are asked to write a feasibility report to introduce a new production method in a large company. This production method would entail a drastic reduction of the taskforce as well as a substantial economic investment. You are addressing the report to the manager of the company. With the help of the checklist below, define the profile of the intended audience of your report.

AUDIENCE CHECKLIST

When writing documents, you should ensure that they satisfy your readers' needs. The following checklist may help you obtain useful information about your audience in order to design its profile so that you can successfully target your document.

1. Who is going to read the document?
   Primary audience
   Secondary audience
   Mixed audience

2. What is the audience's level of understanding?
   Does the audience have a high, low or lay level of the subject matter?
   Should technical terms be defined?
   Is background information necessary?
   Would graphics help understanding?

3. Why does the audience need the document? What action is expected from the reader?

4. What format would best suit the audience's needs and the purpose of the document?

5. What is the audience's position with respect to the organization and the writer?
   Is the reader occupying a high or low position?
   Is the reader a peer, a client or a student?
   Does the reader work for the writer or vice versa?
   How formal should the writer be, considering the reader's position and their relationship?

6. What may the audience's attitude be towards the subject/message?
   Positive   Negative   Neutral (e.g. uninformed, uncommitted)

7. What are the main traits of the writing situation to be considered?
   What is the organization's stance on the subject?
   How much information should/can be included?
   Under what circumstances did the need for the document arise?
   How important is the document for the company?
   What external groups may have an influence on the document?
   How much power does the reader have? Does he/she have enough authority to make a decision?

---

8. How will the audience read the document?
   Will they skim, scan, or do extensive reading?
   Will they read the whole document or parts of it?
   Will they read receptively or critically?

9. How do you expect the audience to react to the document?
   Immediately or slowly
   Enthusiastically
   Receptively
   Negatively

2.3 Analyzing purpose

Technical documents are usually written to satisfy a requirement and consequently to enable the reader to act. Technical documents, then, usually have a purpose. The following questions may help you determine the purpose of the document you are going to write.

What is the purpose of this document?
What do I want to achieve with this document?
What action do I want the reader to take?
What use may the reader make of the document?

When dealing with the purpose, you must remember that although documents actually have a general or primary purpose, they may also have multiple/secondary purposes due to the readers' specific needs; not all readers have the same purpose when reading the same document. For example, while the primary goal of a lab manual is to give instructions on how to use a given device, it can also serve to provide information about the different components of that device. Another example would be a leaflet that describes and informs readers about the latest improvements on a particular product (primary purpose). At the same time, obviously, the leaflet may also be trying to persuade the reader to buy this improved product (secondary or underlying purpose).

Some general purposes found in technical writing are:

To recommend. The writer suggests the reader should undertake a specific action. Documents that include this purpose are: proposals, feasibility and recommendation reports.

To instruct. The writer tells the reader how to do something and why it should be done. Documents that are used to instruct are: technical user's manuals, instructional procedures and process descriptions.

To persuade. The writer tries to convince the reader to take a specific action. In order to convince the reader, the writer must give enough evidence through well-supported arguments that the situation requires action. Documents with a persuasive purpose include: resumes and cover letters, grant applications and technical advertisements.

To report. The writer gives an account of something seen, heard or done, which is frequently followed by an action. This kind of technical writing mainly includes types of reports, such as progress reports and periodic reports.

To inform. The writer provides information to the reader who wants and needs to know about scientific and technical data but has no intention of taking action or making a decision. Documents that primarily inform include: science articles in popular magazines, technical periodicals, catalogues, brochures, bulletins, descriptions, literature reviews and textbooks[1].

The purpose of most technical documents is self-evident and explicit (e.g. user's manuals, resumes, leisure science articles), but sometimes you may need to overtly state the purpose of the document. In these particular cases, it may be useful to include a *purpose statement* at the beginning of the document. This *purpose statement* will be the answer to the question "what is the purpose of your document?" However, it is not advisable to write "the purpose of this document is...", as it is considered a wordy cliché. When it comes to writing a purpose statement, try to make the statement more direct and dynamic by incorporating an action verb such as *demonstrate, develop, conduct, summarize, illustrate, suggest, conclude, classify,* etc. Note how the purpose cliché below may be improved.

Cliché purpose statement:
The purpose of this memo is to define the behavioural characteristics of interoperability requirements for Internet firewalls.

Improved purpose statement:
This memo defines the behavioural characteristics of interoperability requirements for Internet firewalls.

COLLABORATIVE TASKS

2-9 Read the following extracts and decide on their intended purpose justifying your choice. Check your answer with your partner.

---

[1] Note that in the case of textbooks, the writers' purpose is to inform but in a didactic way.

Sample A
Dear Sirs,
We have seen your advertisement in The Metal Magazine, and would be grateful if you could kindly send us details of your chromium-plated handles.
Please quote us for the supply of the items listed on the enclosed enquiry form, (…)

Sample B
This textbook presents an introduction to electrical communication systems, including analysis methods, design principles, and hardware considerations. We begin here with a descriptive overview that establishes a perspective for the chapters that follow. (…)

Sample C
In this study, the annual sales, experience, markets, growth strategies, technology and leadership of 36 recycling firms were evaluated to determine the characteristics that helped these companies survive in this plastic recycling industry. (…)

2-10 Read the following sentences and rewrite them avoiding the cliché purpose statement. Check your answer with your partner.

In keeping with this special issue on biometrics, the purpose of this paper is to present the facilities and network access-control applications of speaker-recognition. This paper focuses on the three applications below.

INDIVIDUAL TASKS

2-11 Read the following texts, identify audience and purpose and find enough evidence to justify your answer. Remember to pay special attention to the title as it may help you in making the decision.

Watching the Nanotube

Now that plasma televisions are here, their makers would have you believe the quest for the ultimate TV is over. After all, these big, flat screens are dazzlingly bright and have a wide viewing angle. They can be hung on a wall or even built right into it. What more could you want?

Well, for starters, how about a TV set that doesn't consume as much power as a toaster oven? For that matter, you would think that any TV technology worthy of the term "ultimate" would be free of significant flaws, which lower-end plasma screens are not. For example, many models costing less that about US $5000 have a distracting tendency to render pure black with a greenish cast.

For reasons like those, bands of researchers in the United States, Europe, and Asia are insisting that the last word in TVs won't be plasma, but rather nanotubes. These exotic molecules of carbon, only a few nanometers wide and perhaps a micrometer long, are at the heart of a new class of big, bright experimental displays that could overcome the power and image quality problems of plasma screens while retaining their brightness and size.

At stake is the richest consumer electronics category in the world: in the United States alone this year, people will buy at least 30 million analog and digital television sets worth more than $12 billion, according to the Consumer Electronics Association (CEA) in Arlington, Va. Digital and flat models—precisely the category targeted by the emerging nanotube technology—are the fastest-growing category, the CEA says.

It will be the first consumer application in microelectronics for these sheets of carbon atoms seamlessly wrapped into infinitesimal cylinders. They have been proposed as the basis for a whole host of technologies, including hydrogen storage, interconnects for chips with ultradense components, and a new breed of transistor.

They're also breathing new life into an old idea—displays based on the phenomenon of field emission. Unlike the liquid displays common in laptops and small video devices, field-emission displays can offer wide viewing angles, and they are inherently less power-hungry than plasma displays, making them cheaper to operate.

With advantages like those, it's no wonder that companies such as Motorola Inc. (Schaumburg, Ill.) and Samsung Group (Seoul, South Korea) are aggressively pursuing field-emission display technology using nanotubes. Samsung, for example, has already demonstrated a full-color 38-inch field-emission display capable of handling normal video frame rates. What's more, a Japanese government-funded consortium was announced earlier this year to develop similar displays, and Sony Corp. (Tokyo) is developing its own nanotube display technology as well.

Plasma demands considerably more electricity than regular television cathode ray tubes (CRTs). A 38-inch color CRT consumes approximately 70 W. A similarly sized plasma display consumes some 700 W—a level of power consumption normally seen only in home appliances, like vacuum cleaners, that are typically in use for only a few minutes a day. Apart from the impact on consumers' wallets, if plasma technology became commonplace, it would result in significant implications for electricity generation and distribution, given that most people watch television for several hours a day and homes (at least in the West) often have multiple televisions. However, a 38-inch field-emission display should be able to provide the same performance as a plasma display while consuming only 50 to 70 W.

Facing a Difficult Problem

From: Thomas Goodman
To: Mr David Williamson (Distronics marketing consultor)

Date: May 31st 2007
Subject: Proposal to reduce losses

As the production manager and only engineer of Distronics, the firm where I work, I am facing a very difficult problem. I think it's high time an important managerial, strategic decision was made because Distronics is on the verge of failure.

As you know, it's about fifty years since the founder of the company decided to create his own company in order to exploit a novel development in MOS circuit integrated technology. Thanks to this, Distronics gained expertise in long-life calculator batteries and started to produce hand-held calculators, a product line responsible for the company's growth during the last few years.

However, nowadays our product line of hand-held calculators is no longer competitive. In fact, an analysis of the current state of the electronics sector and of Distronics, which has recently incurred heavy losses, calls for a good decision to be quickly made at this turning point.

As I see it, Distronics has at least four alternatives:

i) we may reduce costs by buying some of the components in China, but keeping the research and development department in Rubí

ii) we may find a partner willing to inject badly needed capital

iii) we may close down the production plant in Rubí and move to China

iv) we may write off calculator production and return to the basic semiconductor component production, which has always been profitable

I strongly believe we can compete successfully in the calculator industry but I'm an electronics engineer who doesn't know much about great decisions and wants to be well-informed of the feasibility of each option before speaking to the founder. I therefore suggest we should meet some day next week. I understand you'll need to gather as much information as possible to be able to analyse the different scenarios and recommend the best alternative for Distronics.

Adapted from Abell, D. & J. Mammond, (1979) *Strategic Market Planning.Problems and Analytical Approaches*. New Jersey, Englewood Cliffs: Prentice-Hall,134-135. Reprinted with permission of the publisher.

2-12 Imagine the company you work for asks you to attend a trade fair of your sector. On your return you are asked to report on the most appealing novelties and innovations you saw and the conclusions reached.

CRITICAL THINKING

2-13 As explained above, technical documents are written to enable the reader to act and therefore may have a general or primary purpose. Can you think of any other element closely related to this primary purpose which has a direct influence on the creation of the document?

2-14 When writing a persuasive leaflet with the purpose of selling a product to a high-tech audience, would you only mention its advantages? Or would you first acknowledge some of the disadvantages of the product, minimizing them, and then highlight the advantages? Justify your answer.

2.4 Considering style and tone

Having considered audience and purpose, we will deal with *tone* and *style* together, as they are very closely related. In fact, the latter are usually determined by the former. This means the writer should choose the appropriate tone and style to accommodate the document to the intended audience and purpose. For example, a memo to be sent to the general manager of the European headquarters, a memo written to a colleague in your department, an application letter, or a leaflet for a potential client is not going to require the same tone and style. Each is a different document with a different audience and purpose, so for the message to be transmitted efficiently it should be written in an appropriate tone and style. Choosing these elements is a decision to be made at this pre-writing stage so you will obtain your focus right from the very beginning.

The most common types of tone and style in technical writing are coherent with the definition of technical writing. If you remember the characteristics of technical communication, good technical writing is *clear* and *concise*. It is *clear* because it is to the point; the writer manages to write with clarity and efficiency by avoiding unnecessary and repetitive words that add nothing to meaning. Technical writing is also *concise* because ideas have to be expressed as exactly as possible; the writer can avoid vagueness by choosing specific and concrete words. Whatever the writer's choice of style and tone, technical prose should not add complexity to the already complex material; on the contrary, if the prose is clear and concise, the writer will allow readers to concentrate on the content being conveyed, not on how it is presented. We also mentioned that the main objective of technical prose is to transmit information not only clearly but also efficiently so that anybody reading a text for the first time easily understands its content and is, at least, convinced that he/she should take the writer's claim or opinion into consideration. This means that as a writer of technical documentation in English, you are likely to need:

- awareness of the different types of tone and style in the English language
- ability to choose the appropriate type of tone and style to be adopted
- mastery of how to use them in your writing.

Both tone and style are mainly concerned with the choice of certain linguistic forms that result from the writer's attitude and his/her relationship to the reader, to the topic, and to the purpose of communication. However, tone and style are expressed in different ways, for example, by word choice, by syntax, by grammar, by idiomatic expressions, etc. They can also be used in combination or independently; that is, a formal style usually goes together with an impersonal and objective tone. Tone and style being different, the differences between them must be highlighted.

Style

Style refers to the way the message is expressed and is determined by the choice of grammar, syntax, vocabulary and idiomatic expressions. The three main varieties of style that emerge from this choice are: *formal, informal* and *slang*[2]. It is important to bear in mind, however, that it is not always possible to draw a clear-cut line between them and that there is a neutral or unmarked variety of style. To be more precise, then, the scale below could be said to represent the different gradients style can range from (Quirk et al, 1987: 25-7):

VERY FORMAL > FORMAL > NEUTRAL   >   INFORMAL > VERY INFORMAL (SLANG)

The following examples may help you clarify the differences:

VERY FORMAL
Pursuant to your orders, it is with great regret that we hereby inform you that your company's shares have been observed to show a significant decline in the stock exchange.

FORMAL
We regret to inform you that the value of your shares has significantly decreased in the last few days.

NEUTRAL
We'd like to tell you that your shares have gone down in the last few days.

INFORMAL
We are sorry to tell you that your shares have just collapsed.

VERY INFORMAL
I dunno how to break the news to you. I've kept my eyes peeled on the shares for a while and the investment has been a complete disaster, sorry 'bout that.

---

[2] Slang is usually defined as words, expressions and their meanings commonly used when talking with friends or colleagues in very informal situations. As slang is not regarded appropriate in technical communication, we will not deal with it in this book.

Formal style usually creates a distance between writer and reader as it tends to reflect a respectful and/or detached stance on the part of the writer; that's why many times a formal style is used when a large audience or a superior is addressed. *Formal style* is essentially characterized by a careful choice of words, a careful use of punctuation and the use of formal language, sentences and expressions that are syntactically and grammatically correct. Even though most current handbooks on technical communication encourage scientists and technicians to adopt a more direct style, academic texts have traditionally been written in an excessively formal style as most scientists seem to be reluctant to incorporate a more plain style when writing.

Informal style, on the other hand, is generally used when the document is aimed at peers and colleagues or when the purpose of the document allows a closer relationship with the reader. It is usually associated with everyday speech and consists of vague and imprecise expressions, a lax use of grammar, syntax and punctuation, and a widespread use of popular idioms, abbreviations and contractions.

The table below summarizes the main differences between formal and informal style. Note that the table contains some general guidelines that should not be taken as prescriptive since you may find variations among writers. There are two factors that may contribute to this variation: writers' idiosyncrasy and the fact that there is a considerable number of technical writers that are non-native. This implies that in some pieces of technical writing we may find an overuse of formal expressions while in others the use of these expressions may be scarce.

	Formal style	Informal style
Word choice	<ul><li>No contractions[3] (e.g. *does not, have not, did not*).</li><li>No conversational discourse markers (e.g. *er, um, well, y'know*) are used.</li><li>Use of formal language and verbs from a Latin origin (e.g. *conduct, discover, investigate*).</li><li>In a very formal style, "legalese" or "bureaucratese" words and expressions are used (e.g. *herein, whereby, heretofore, attached hitherto, inasmuch as, enclosed herewith, keep me timely advised, aforementioned*).</li></ul>	<ul><li>Use of contractions</li><li>Use of colloquial or even slang expressions (e.g. *guy, pretty good*), contractions (e.g. *'cos, gonna, wanna*).</li><li>Use of imprecise expressions (e.g. *sort of, kind of, like*).</li><li>Frequent use of familiar language, phrasal or prepositional verbs (e.g. *carry out, find out, look into*).</li></ul>

---

[3] Nowadays contractions have begun to be used discriminatingly in formal writing.

Syntax	▪ Longer and well-written sentences, strictly following syntactic rules. ▪ Use of formal transition markers (e.g. *nevertheless, despite, on account of*). ▪ Passive voice and impersonal structures are frequent.	▪ Short, more direct and less elaborated sentences ▪ Use of coordinators or simpler transition markers (e.g. *and, so, but*). ▪ Active voice and the use of personal pronouns are more frequent.
Grammar and punctuation	▪ Conventional rules of grammar and punctuation are respected.	▪ Punctuation and grammar are more lax and correctness is often disregarded.

Table 2.1 A comparison of formal and informal style

Notice how the paragraphs below exemplify the differences between formal and informal style as outlined above.

Formal style sample

In the previous section, we derived results for generating momentum in desired directions. In this section, we explore the implication of this for steering, using an analogy to wheeled mobile robots. While the results are fairly straightforward applications of the above algorithms to motion planning, they are useful because they allow us to control the robot over a non-local region of the state space. In other words, instead of being forced to patch together purely local results as is generally done in steering algorithms for non-holonomic systems using cyclic inputs (Cortes et al. 2000), we can determine motion plans for more non-local control goals.

Reproduced with permission from Melsaak, K and Ostrowski, J. (2003) Dynamic robotic locomotion systems. *The International Journal of Robotics Research* Vol. 22 (2), 87. (© Sage Publications 2003) by permission of Sage Publications Ltd.

Informal style sample

It's 1960 or so, and you're a 12-year old living in a suburb somewhere on the East Coast. It's late on a school night and you're in bed, but you're not sleeping; you're listening to a big old Atwater Kent radio that you inherited when your mother got one of those new transistor radios for the kitchen. Tonight you've tuned to WOR in New York at 7 10 AM, and you're listening to this guy named Jim Shepherd who's telling a funny story about when he was a kid and blew up his ham radio one day, scaring his mother and almost burning down the house. Then he starts talking about the radio and vacuum tubes, and he says that those new things called transistors may be good for some stuff, but they'll never replace tubes altogether: "When you get down to putting out 50,000 watts of radio frequency, no sir. You gotta call in the big boys with the fans blowing on 'em. With the water running through 'em to keep 'em cool".

Source: Wolverton M. (2002). The Tube is Dead. Long Live the Tube. The vacuum tube has been obsolete for decades-- and here's to stay. *Invention and Technology*. American Heritage. Fall 2002, Vol.18(2),.28-32. Reprinted by permission of American Heritage.

INDIVIDUAL TASKS

2-15 Considering that style has different gradients of formality, read the following series of near-synonyms and order them according to their degree of formality. Try numbering the series from formal to informal (1 very formal, 4 informal).

terminate, stop, cut off, interrupt

give a buzz, telephone, ring up, make a call

depart, do a runner, fly, run away

breakdown, collapse, crackup, failure

obtain, acquire, get hold of, score

find by chance, run across, encounter accidentally, bump into

later, afterwards, subsequently, thereafter

team up, work together, collaborate, yoke

cut down, reduce, diminish, abridge

exhaust, finish, use up, deplete

wind up, finish, conclude, cease,

be in the know of, know, be cognizant of, be aware of

2-16 The sentences below have been written with informal verbs. Rewrite them using formal synonymous verbs:

1. Do not disturb John; he's now concentrated on **working out** a mathematical problem.
2. My application for the internship in that audit-consulting company was **turned down** because, they said, I do not have enough experience in financial software.
3. The accountant assistant made a mistake when **making out** the cheque.

4. The After-Sales Department promised they would **look into** our complaint and suggest a solution.
5. I **put forward** the idea of introducing a new payment system to keep the personnel more motivated, as the rate of absenteeism has been increasing lately. The Executive Board thought this suggestion was very good and accepted the proposal.
6. The alarm **went off** when some burglars broke into the office building during the weekend.
7. After the terrorist attack, the manager decided to **lay off** a hundred workers because business was very bad.
8. The company's plan to establish a sister company in Spain **fell through** because the tax governmental policy is not appealing enough.
9. We have **broken off** negotiations with this potential supplier.
10. The pipes in the system were leaking and the workshop manager **got round** the problem by welding the loose joints.

2-17 Read the following sentences and identify the informal structure or expression(s). Then supply more formal words or structures:

1. They got good results and announced that they'd repeat the experiment with a different kind of method.
2. The reply to our proposal was sort of negative, but one of the things they said would happen was that our sales would skyrocket for some time.
3. Right, this year's benefits have been pretty bad, but we'd better wait and see, 'cos the political and economic situation of the country is mighty nice.
4. The new guy in the Quality Department has messed everything up because he didn't have clear task guidelines.
5. The factory workers seem to be chuffed about the working conditions for the time being.
6. We'll have to get in touch with our suppliers to work out what happened with those defective spares they sent.
7. The Executive Board will hit the ceiling as soon as the news that such an important client has been lost is known. However, the company shouldn't dwell on this news because there are other important problems that need a solution right now.
8. The sales department can't put up with delays in the assembly line as this only benefits the company's competitors.
9. This new stuff has come in handy for our research laboratory.
10. The purpose of this report is to present a down-to-earth description of the up-to-date developments in the field of water supply and sewerage.

Tone

On the other hand, *tone* reveals the writer's attitude towards the subject and the reader and it is mainly determined by the choice of words and content. There is a wide range of tones a writer may adopt:

> *personal, distant, impersonal, ironic, business-like, sarcastic, stuffy, polite, casual, tentative, insulting, assertive, insistent, condescending*

There is a great variety of tone types but the most commonly found in technical writing are the following:

Personal. This tone is sometimes also known as the 'you-approach'. It basically reflects that the writer is not keeping a distance between the readers and himself, remaining aloof or detached from the audience; instead, writers adopting such a personal stance usually address their audience, trying to get them involved, and appear to be more personally committed to their document. With this tone writers are more direct and manage to make their texts more dynamic and agile—sometimes even more persuasive. This kind of tone is typically characterized by:

- a frequent use of personal pronouns
  e.g. I, we, you

- use of active voice (rather than passive)
  e.g. Bell invented the telephone (rather than *The telephone was invented by Bell*)

- the use of expressions that refer to the reader and to the relationship between the reader and the text, even including direct questions to the reader
  e.g. This chapter will help you decide…
    The role of this brief introductory chapter is to present a few terms and concepts to guide us into the body of the text.
    Don't you think that such an exposure to radiation may be dangerous for your health?

- the type of content. The use of somewhat personal anecdotes (i.e. short passages which are usually narrated in active voice, first person, and in a more informal style), a short commentary on the writer's own limitations, or questions to introduce a new topic or idea and which can be written in the first or the second person also favour a more personal tone.
  e.g. As an example, in the laboratory where I was employed, we had a problem with a cylinder head and…

The *didactic tone* is also usually associated with the personal tone. This didactic tone is the one that a writer may adopt in textbooks, for example, when making a direct appeal to students' needs and interests, when commenting on the difficulty or simplicity of the content to be explained, or when justifying why a given concept needs to be prefaced by a theoretical framework. In the didactic tone, it is frequent to find the "we" and "you" pronouns and imperatives as they reflect that the expert writer is attempting to get closer to the reader. Notice these aspects in the examples below:

> Note that you won't usually be able to identify such a problem if you've had no prior experience with it.

> In picking preload or torque for such a joint we should answer two questions: how much initial clamping force do we want in this joint, considering the service loads and conditions the joint will face? How much clamping force—and scatter in clamping force—can we expect from the assembly torques, tools, procedures, etc. we plan to use? Let's see how we might answer these questions in an economically, acceptable way.

Copyright (© 1998) From *Handbook of Bolts and Bolted Joints by* Bickford, J & S. Nassar. New York: Marcel Dekker Inc., 659. Reproduced by permission of Routledge & Francis Group, LLC.

Impersonal. With this tone the writer remains in the background, avoiding any direct and personal mention of him/herself. The impersonal tone can be achieved by means of the passive voice, "introductory *it*" statements, and certain formal expressions that writers use to keep a distance with the reader. This is because the focus of impersonal technical writing is on the object of study rather than on the doer of the action. The use of all these impersonal expressions makes the text more lengthy and bulky, and its abuse may diminish the readability of the document. Impersonal tone is usually characterized by:

- a frequent combination of impersonal statements (mainly in the passive voice) and formal expressions

  e.g. The behaviour of the materials in an engineering design can be said to relate directly to mechanisms occurring at those minute levels. An understanding of this relationship allows the proper material to be selected for a given design.

- the use of expressions that refer to the text (its purpose, presentation and organization)

  e.g. The purpose of this report is…
  In this chapter the relations between strength and static loading are considered.
  This chapter focuses on water supply systems. First planning, design and operation of the collection will be dealt with and then purification transmission and distribution works will be described.

- the strictly scientific and technical content. The information is to-the-point and closely related to the subject matter; in other words, there is no room for personal comments and anecdotes.

  e.g. Steady progress is being made in photovoltaics (PV) for terrestrial power. Production volumes have been rising nearly 20 percent annually, costs have been falling, the range of products expanding, and efficiency improving, while readability warranties now cover up to 20 years.

The example below illustrates these traits of impersonal tone and formal style:

> One of the first design tasks is determining appropriate movements of the wipers. The movements must be sufficient to ensure that critical portions of the windshield are cleared. Exhaustive statistical studies reveal the view ranges of different drivers. This information sets guidelines for the required movements of the wipers. Fundamental decisions must be made on whether a tandem or opposed wipe pattern better fits the vehicle. Other decisions include the amount of driver and passenger side wipe angles and the location of pivots. Figure 1.1 illustrates a design concept, incorporating an opposed wiper movement pattern.
>
> Once the desired movement has been established, an assembly of components must be configured to produce the wipe pattern. Subsequent tasks include analyzing other motion issues such as timing of the wipers and whipping tendencies. For this wiper system, like most machines, understanding and analyzing the motion is necessary for proper operation.

Source: Myszka, D. H. (2002). *Machines and Mechanisms. Applied Kinematic Analysis*. (2nd ed.). New Jersey, Englewood Cliffs: Prentice-Hall, 1. Reprinted with permission of the publisher.

This text is a good example of an impersonal tone as it contains passive and impersonal sentences and omits the use of personal pronouns. However, in terms of degree of formality it would be on a lower scale compared to the formal text on page 59 because there are only a couple of complex sentences and connective expressions are barely used. Note, on the other hand, that the word choice is reasonably formal as the author uses words such as *sufficient, exhaustive, illustrate* and *subsequent*.

Tentative. A common feature of technical and academic English is the need to be cautious because quite often the writer is not completely certain about the truth-condition of a statement. With this in mind, we can define the tentative tone by explaining what it is not: it is the opposite of assertiveness. When you need to be careful with the information you want to transmit you can adopt a *tentative* tone that will convey caution and lack of certainty. This tone may be achieved by means of the use of expressions of probability such as *perhaps, likely, apparently, possibly, probably*, modal verbs like *appear, seem, could, may, might*, approximate expressions like *roughly, somewhat, approximately*, etc. and subordinate constructions like *that-clauses* and *passive-infinitive structures*. Since technical and scientific writers very often want their readers to accept their claims, arguments and counter-arguments, such expressions can also contribute to showing the writer's respect for the audience. The passage below has been written in a tentative tone:

> The analysis of pre-Poll data presented here **suggests** that much of the wide variance in the willingness to pay a premium for renewable energy and energy efficiency among the general public can be explained by demographic factors, such as age, salary, and education. The Polls led to much greater interest in relying upon energy efficiency as a «first-choice» energy option. The concept of using energy efficiency as a resource to a utility system **may** have been unfamiliar to some respondents prior to the Poll, and the information presented at the Polls **may** have introduced participants to an intriguing concept. (...) While **it is doubtful** that the high level of support for renewable energy and energy efficiency **suggested** in these survey responses will lead to similarly high levels of participation in voluntary programs, such survey results **may** provide **some** guidance to utility resource planning activities and public policy formulation.
>
> Reprinted from Consumer demand for 'green power' and energy efficiency. *Energy Policy.* Vol. 31(15) by Zarnikav, J., (1971). © 2003, with permission from Elsevier.

Note how the degree of certainty and assertiveness in the same passage can change just by modifying a few expressions.

> The analysis of pre-Poll data presented here **indicates** that much of the wide variance in the willingness to pay a premium for renewable energy and energy efficiency among the general public can be explained by demographic factors, such as age, salary, and education. The Polls led to much greater interest in relying upon energy efficiency as a «first-choice» energy option. The concept of using energy efficiency as a resource to a utility system **was certainly** unfamiliar to some respondents prior to the Poll, and the information presented at the Polls **introduced** participants to an intriguing concept. (...) The high level of support for renewable energy and energy efficiency **shown** in these survey responses will lead to similarly high levels of participation in voluntary programs and such survey results **will** provide guidance to utility resource planning activities and public policy formulation.

Finally, let's consider some further points in relation to tone and style. The first aspect is related to possible combinations established between tone and style. *Concordancing* is a useful technique in computer analysis that can give us practical insights into the nature of word combinations and which provides us with samples that can be used for analyzing linguistic features of authentic texts. Below you will find an example of concordance data from two different genres, namely, business letters and personal letters. Can you identify each type? What style features characterize each genre in terms of vocabulary, syntax, grammar and punctuation? (Refer to Table 2.1 in the style section).

```
 1   have avoided a lot of aggravation. However, a substantial portion of our fees
 2      due to lack of work in our area; however, I highly recommend Jim for employm
 3         as the company has grown. However, we have been having problems.
 4                          in my opinion. However, these expenditures should be accou
 5   preciate your attention to detail; however, if we do have branches that are no
 6    to those outside my organization; however, neither was your effort on my beha
 7    was shipped to Lafayette in July; however, the job was unsuccessful due to ma
 8     exceeds the engineering estimate; however, late charges, rental charges, and
 9   nue serving you on a credit basis; however, we need your cooperation.
10   orrect, there is no need to reply; however, if you find a discrepancy, please
11   preciate the quality of your work; however, it has come to my attention that s
12   as always been < fairly relaxed >; however, recent complaints from < customers
13            to avoid this situation. However, we ask that, in an emergency, emp
14   dred) raincoats in this material; however, the market here will not stand the
15   take the matter up with shippers; however, for the present we should be oblig
```

```
 1   iding in him. Now I realize that, however noble my intentions, I was wrong to
 2   spleasure with Best Value. My hope, however, is that you'll make that step unne
 3   t how old he actually is. Recently, however, he's been experiencing shortness o
 4   s prayer. At this stage in my life, however, I'm not convinced that I'm ready t
 5                            I do, however, have one major concern: traffic fl
 6        I was particularly offended, however, when your receptionist telephoned
 7   to life. It never occurred to me, however, that a background in chemistry and
 8           My experience tells me, however, that most people feel worse after
 9   ted by her death. Six months later, however, he began dating an old acquaintanc
10   hat Wednesday trip with me; these, however, are not the reason for the return.
11   shelves anywhere. One store clerk, however, took the time to give me your addr
12   he price was $5.99 six months ago; however, since I do not know the appropriat
13   When I tried to use it last week, however, the power button would not work. I
14   ents immediately. For some reason, however, the first payment was credited as
15   ally left on my own accord. Today, however, I received a $120 bill for "treatm
```

As you may have guessed the first corpus is taken from business letters whereas the second is from personal letters. Business letters tend to be written in a formal style as can be seen in the samples above, where we find formal words and expressions, passive structures and conventional punctuation. On the other hand, the examples from the personal letter corpus reflect a more informal style as they include more familiar words and expressions, and a more lax use of punctuation. Usually associated with these two different styles are two common types of tone. Can you find out what kinds of tone are usually related to formal and informal style?

To sum up, a formal style tends to be accompanied by an impersonal tone whereas a somewhat informal style is frequently used in combination with a more personal, friendly tone. At the same time, writers should try to avoid making an excessive use of extremely formal words and expressions as the text may sound pretentious. In addition, writers of documents such as reports and research articles tend to adopt a somewhat formal style and impersonal tone so that their texts carry the ring of objectivity and credibility.

It is also worth bearing in mind that technical writers should be careful when choosing the informal style and personal tone combination. Documents written in an informal style and personal tone uniformly from beginning to end are very rare. Sometimes different types of tone and style appear in different parts of the same document. For example, a somewhat informal style and personal tone can be found in results and conclusions of certain documents. Similarly, a formal style and a tentative tone in these parts of the document are also used when writers are trying to be cautious. In other parts of the document, for example when a process is described, writers tend to be more impersonal and hence adopt a more formal style.

Finally, by always taking into account your audience and purpose, try to monitor your style and tone so that the message your reader receives is the message you intend to convey. Also remember that in technical communication it is a good idea to adopt a polite and considerate tone throughout the document and avoid being ironic or sarcastic.

INDIVIDUAL TASKS

2-18 Read the following situations in which different kinds of documents have to be written for different audiences. Then for each writing situation indicate the different contextual features and finally decide on the most appropriate tone and style.

SITUATION 1
You are the general executive of an international company and have to write a memo to your Production Managers in the European and North American divisions asking them to analyze the pros and cons of moving production to China.

SITUATION 2
You are one of the production managers in the situation above and have to write a feasibility report to your general manager.

SITUATION 3
The general executive has decided to move the production line in your local sister company to China. As the production manager, you have to write to all your staff about the imminent closing down of your production line.

SITUATION 4
You are the first engineer working in the production department to informally hear the bad news. You decide to quickly email all the other engineers in the factory about the closing down of the production plant.

Contextual features		Situation 1	Situation 2	Situation 3	Situation 4
No. of readers	1 or 2				
	Small group				
	Large group				
Degree of closeness	Known readers				
	Unknown readers				
Status of participants	Boss				
	Peer				
	Subordinate				
Shared background knowledge	Yes				
	No				
Shared topical knowledge	Yes				
	No				
Style					
Tone					

2-19 Read the following letter and the different replies to it. Then from the adjectives in the two groups below, choose those that best describe the kind of tone and style used in every reply and write them down in the space provided. With regard to tone, remember that more than one adjective may be used to describe the replies.

14, Oxford Street
London

February 1ˢᵗ 20XX

For the attention of the general manager

Dear Mr Reynolds,

It's a pleasure for me to inform you about the standard of your after-sales service. As you know, we have been buying several machines from your company for some years now and have even recommended Fox machines to other companies, as your products are excellent.

However, since your company decided to apply the Quality concept to the after-sales service, some changes have been made that directly affect our service. An old computer and a loyal employee were replaced with new hardware and software. In the past, if the speed of processing orders slowed down, the clerk in the department worked longer hours to catch up with the backlog of orders, and enquiries were always answered courteously and efficiently. Now the boy dealing with enquiries and after-sales complaints promptly lays the blame on the production department. When a machine breaks down, we are now told by the after-sales service that 'zero defect' is the current goal in Fox Industries, that poor quality materials are no longer used, and that we must be wrong. If we want one of your service engineers to come to our factory at 48 hours notice—as it was the case before Total Quality Management reached the after-sales service—we now have to speak to your maintenance head and explain the problem again. We are very glad as we have had the chance of meeting other employees in your company. As you know, making new friends is always such a good thing, so we congratulate you.

We look forward to hearing from you.

Yours sincerely,

D. Regal

REPLIES TO THE LETTER

A

Dear Mr Regal

Thank you very much for the letter. Given that a meeting is to be held with the quality department, no definite answer is possible yet. The general management will look into your complaint and your company will be informed of any important decision in due time.

B

Dear Mr Regal

Thank you for your letter but we don't accept your complaint. You must be wrong about our after-sales service, as absolutely nothing of the like has ever happened in our company.

C

Thank you for your letter of February 2006. Our company has just started to implement this new service and some mistakes may have been made during this initial 'trial-and-error' period. I am personally involved and have spoken to several people in the after-sales department. As I understand that our service is currently not being as efficient as it used to be, I suggest that you contact the after-sales department through the email address created for that purpose instead of ringing. For the time being, let me tell you that your company can count on a service engineer at 24 hours' notice for this month. In Fox Industries we are working hard to ensure that this situation is not repeated.

STYLE   very formal, formal, neutral, informal, very informal

TONE   polite, irate, sarcastic, aloof, vigorous, assertive, demanding, empathetic, condescending, cold, humorous, straightforward

2-20 Imagine you have detected a mistake in the registration fees of this academic year. The university has mistakenly charged you 150 Euros more. Write a short email to the person responsible in the accounts area of your faculty, briefly explaining your case and asking for an amendment.

COLLABORATIVE TASKS

2-21 In pairs, rewrite the neutral statement below, providing the remaining gradients of formality (very formal, formal, informal and very informal). Then compare your answers with those of another team.

Neutral statement: This paper puts forward a new theory that will become an important breakthrough in microelectronics.

2-22 Read this e-mail and analyze the writing situation (Who's the writer writing to? What's the writer's purpose?). Determine whether any change in terms of style or tone should be made and then rewrite it.

From:	mike.harris@upc.edu
To:	Noah Miles
Subject:	Technical writing class

Hello,

About the 'Technical written communication' class, I have a compulsory laboratory session next Tuesday at 10:00, so it is impossible to come to do the test. Please change the date from Tuesday morning to Tuesday afternoon.

Thank you

Mike Harris

2-23 As seen in task 2-22 above, the choice of tone and style depends on the writer's purpose. Try to write a text with a very clear purpose and with an inappropriate tone and style. You can also try to find a published text with the same problem. For example, a research article with the purpose of informing of a new finding cannot be written with a casual tone and informal style or a leaflet aimed at persuading cannot be written in a tentative tone.

2-24 Read the following passage written in a formal style and impersonal tone. Then imagine you have to send it to a colleague in your department and adapt it.

It is with great pleasure that I herewith invite you to participate in the Curriculum Project Forum at the Faculty of Engineering. With the aim of broadening participation concomitant with the creation of an academic debate, the faculty board posits it is necessary to elaborate a project inasmuch as it enables the faculty to face the future with an appropriate curriculum. The members of the community willing to collaborate in the elaboration of the project are encouraged to raise relevant concerns, bring up new ideas, and submit proposals. Should you be interested, do not hesitate to write to us, specifying in which of the four work teams that have been created for that purpose you prefer to participate.
Replies should be sent by March 12th.

The Dean.

 C&

2-25 Below is a business letter to a client. Read it and decide whether its style and tone are appropriate. If you consider they are not, rewrite it in pairs.

30, West Hampstead
London

30th January 20XX

Dear Tom,

We were so pleased to hear from you after such a long time! You were a very good client of ours for a long time, but out of the blue you broke off the relationship. What was the reason for this interruption? Were you dissatisfied with our products or service? Anyway, I hope we can begin our relationship again, now that you've contacted us.
As you'll see in the brochure I'm enclosing, we've made several important improvements to our products (design and specifications) and, surprising though it may seem, prices have levelled off. I hope you appreciate this, considering the country's inflation rate over the past two years. I hope you find our products interesting enough (our present product range is super!) to place an order soon.
I'll be expecting your reply.

Best wishes,

Jessica Nelson

CRITICAL THINKING

2-26 What would be the effect of a text that incorporates many tentative expressions?

2.5 Generating ideas

Before beginning to write, you will probably need to gather information and ideas for your document. Although this may seem a triviality, it is not, even if you have very clear ideas about what to write. Gathering information and ideas is an important pre-writing step that will determine the contents and also the quality of your text. No matter how well you write, if your document is not complete, that is, if it does not include all the information it should include, it will not be effective. Moreover, this preliminary step will help you overcome blank page syndrome, get started and as such initiate the sometimes inhibiting writing process. Additionally, this idea-generating process should not be abandoned in later stages as it may continue to be helpful to further develop points.

One of the first pre-writing tasks you will have to face is the gathering of data if you wish to write with authority on the topic. Consequently, you may have to research different sources and collect as much information as possible so as to be able to support the facts you expound. The type of research you carry out will vary depending on the type of document you plan to write and its purpose, but you should try to collect as many useful details as you can in order to communicate your content effectively to your audience. When collecting information remember to cite the sources your arguments are based on as copying from other sources without citing them is plagiarism. This does not conform to the ethics of writing in general and would make your technical document inaccurate, questioning your credibility.

As well as information gathering, there are also other techniques that have proved useful in this initial stage of generating ideas. Some of the most helpful pre-writing techniques are the following:

Brainstorming/Listing

This is an activity that can be performed individually or even better in groups. It consists in letting ideas flow quickly without inhibitions so as to collect as many suggestions as possible. The ideas will mainly be related to the content of the document but may also include some organizational aspects. Here is an example of a brainstorming list on the topic *Applications of the Internet*.

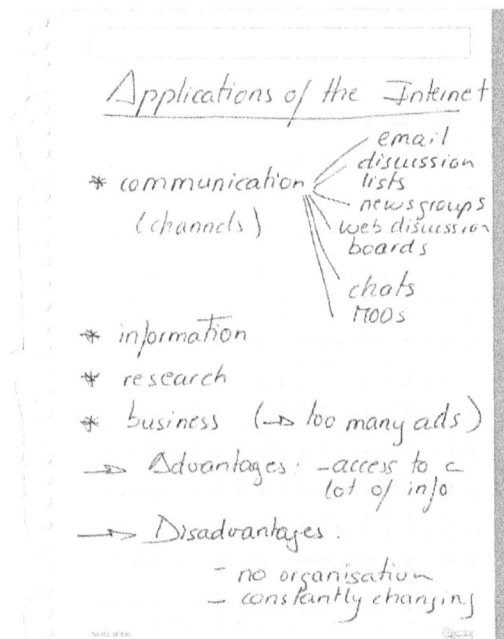

Fig. 2.1

Below you will find useful tips to bear in mind when brainstorming:

- Jot down everything, even ideas that seem poor or obvious. The more you write, the more ideas you will have to choose from, and somehow you may even manage to use ones you were going to abandon initially. If some ideas are clearly off-topic, they can be removed at the end of the activity.
- Don't judge the quality of ideas at this stage. The brainstorming activity should be as free as possible, without any kind of censorship and odd or unconventional ideas should also be accepted. Any attempt to assess or to organize brainstorming will hinder creativity and so limit the possibilities of this technique.
- Don't worry about writing good and complete sentences and don't check grammar and spelling. Imagine the paper is the prolongation of your mind and write quickly as your ideas flow. Simply note down fragments or key words that will later remind you of that particular idea.
- Use arrows, circles, boxes or any drawing you may find useful to connect the ideas in your list. Also remember that this list is for your eyes only, so you do not have to present it neatly.

Clustering/Mapping

To begin clustering you simply have to write your topic at the centre and then radiating from it, like the spokes of a wheel, write down subideas associated with the main topic. If you associate new ideas, supporting points and examples to a secondary idea, cluster them around it. You can also even create different clusters, which will provide you with a variety of perspectives on your topic. This technique can also be used at the end of a brainstorming session in order to organize the different ideas generated there. The figure below is a mind map used to organize the ideas of the brainstorming list shown in Figure 2.1.

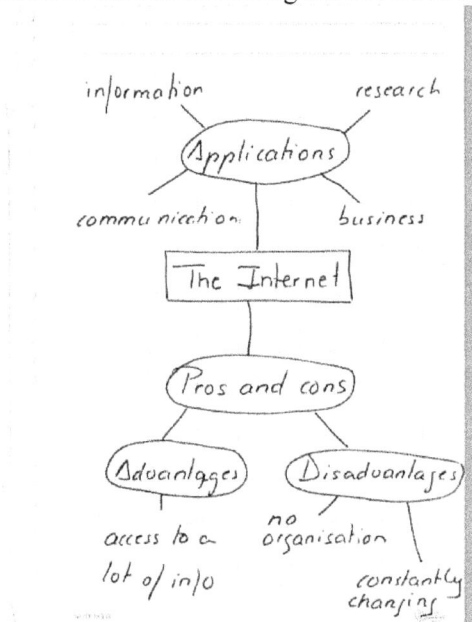

Fig. 2.2

Using questions

Another way to obtain ideas is by asking questions and trying to answer them. This technique will offer you enough data to focus and develop your topic. A typical example is derived from newspaper articles in which reporters usually answer six basic questions: *Who? What? Where? When? Why? How?* These six aspects are essential in any topic, although they may be expanded by adding other questions, so you can use them to explore any issue you intend to write about. Another option is to use another series of questions so as to encourage creativity and thinking. This may provide you with a variety of ideas and may promote an interesting and valuable discussion if the activity is done cooperatively.

The following extensive list of questions, adapted from Jacqueline Berke's *Twenty Questions for the Writer*[4], covers a variety of functions or ways of looking at your topic. Not all of them may be appropriate but they will make you reflect on your topic and may provide you with useful insights into it.

1. What does X mean? (Definition)
2. What are the various features of X? (Description)
3. What are the component parts of X? (Simple Analysis)
4. How is X made or done? (Process Analysis)
5. How should X be made or done? (Directional Analysis)
6. What is the essential function of X? (Functional Analysis)
7. What are the causes of X? (Causal Analysis)
8. What are the consequences of X? (Causal Analysis)
9. What are the types of X? (Classification)
10. How is X like or unlike Y? (Comparison)
11. What is the present status of X? (Comparison)
12. What is the significance of X? (Interpretation)
13. What are the facts about X? (Reportage)
14. How did X happen? (Narration)
15. What kind of person is X? (Characterization/Profile)
16. What is my personal response to X? (Reflection)
17. What is my memory of X? (Reminiscence)
18. What is the value of X? (Evaluation)
19. What are the essential major points or features of X? (Summary)
20. What case can be made for or against X? (Persuasion)

Cubing

This is a technique intended to make you explore your topic from six different angles. One way of applying it is by setting a time for writing about six aspects of your subject, devoting an equal amount of time to each. This will provide you with six different ways of looking at your topic, allowing you to choose the one that is most interesting. Within these views you

---

[4] From: Berke, J. & Woodland, R.. (1994). *Twenty questions for the writer: A rhetoric with readings.* (6th ed.). Heinle & Heinle Publishers.

may also find a new perspective you hadn't considered before and even if some of these perspectives are not used in the actual writing, they may enrich your text and give more strength to your thesis. Another way of applying this technique is by analysing a topic from the following six predetermined facets:

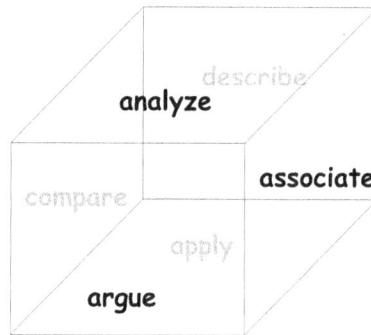

Describe. Elicit some of the most important characteristics and aspects of the topic.
Analyze. Provide some further details.
Apply. Mention some possible uses and applications.
Argue. Provide reasons for or against.
Compare. Compare it to other facts or objects.
Associate. Show or establish possible connections to other topics or aspects of the same topic.

Another similar technique consists in exploring the topic by using Aristotle's classical categories. Instead of six perspectives, as in the cube, there are only the five approaches described below:

Definition. Define the topic and mention its possible constituents, also provide a classification if appropriate.
Comparison. Describe possible differences and similarities and the way the topic differs from other concepts or things.
Relationship. Point out causes and effects, antecedents and possible contradictions and opposition to the topic.
Circumstance. Set forth the topic's possibilities, past facts and future predictions.
Testimony. Make reference to authorities in the field, statistics, theories, precedents and personal experience to provide support to the facts.

Freewriting
Freewriting consists in letting your thoughts on the topic flow freely, writing them down in a paragraph at the same time. You should concentrate on content rather than on form and

consider quantity more than quality. It is also important to omit any sort of censorship, not to stop writing and not to read over what you have written during the entire activity. For this it is a good idea to allot a fixed time, for example ten minutes, to the task. Like brainstorming, this technique is very useful not only to generate ideas but also to get started. The resulting text may be full of fragments, unfinished sentences, grammatical mistakes, spelling errors, incoherencies, etc. but it constitutes excellent material to be developed and rewritten to help you obtain an effective final version.

Note-taking
Making notes as an idea-generating technique may be considered a humble variation of writing a diary. Often good ideas may come to your mind when you are on the bus, in a meeting, in the street or somewhere you cannot write them down. For this reason it is a good idea to carry with you a small note-book, a piece of paper or anything you can think of to record these ideas so that you can use them later. You can write down whole ideas, details, examples or simply words that you may like to use later in your text. Again, since this activity is primarily concerned with generating ideas, you shouldn't worry about language correctness at this point.

COLLABORATIVE TASKS

2-27 Write the topic *Alternative Sources of Energy* at the top of your paper. Next, individually or in groups, brainstorm writing down any ideas related to it that come to your mind. Remember not to assess the ideas you are writing and not to worry about grammar and spelling at this point. Once you finish the activity, read over your list and use arrows, circles and boxes to highlight or to connect important points or ideas. Finally remove from the list any ideas that are clearly off-topic.

2-28 Also on the topic *Alternative Sources of Energy* and using the improved list of ideas generated in the brainstorming session in task 2-26, try clustering. In the centre of your paper write the topic and then make associations branching out from this centre. Once you have organized all the ideas in the list you may want to begin a new cluster in order to get a new perspective on the topic.

2-29 On the topic *The Use of New Technologies in the Information Age*:

a) Individually write down five questions. Try to ask questions that prompt ideas that may interest potential readers. Then, in small groups compare your questions and exchange ideas. Finally, decide on a final list to share with the rest of the class.

b) Now answer as many questions as possible from Berke's list in order to see how far they help you widen your previous list of ideas.

INDIVIDUAL TASKS

2-30 Apply the cube technique to explore the topic *"Mobile phone technology nowadays"* from six different perspectives. Once you finish this task, try to approach the topic by using Aristotle's classical categories. Finally compare both versions in order to spot any similarities or differences.

2-31 Choose a topic related to your field of studies. Think about it for a couple of minutes. Now freewrite for ten minutes, letting all your ideas flow into a paragraph without any censorships or judgement and without worrying about coherence, correctness or sentence structure. As soon as the ten minutes are over, stop, read over what you've written and assess the different ideas on the paper.

2.6 Outlining

In the gathering information and idea-generating process, a considerable number of ideas may have arisen which is likely to need some filtering and grouping. However, you need some criterion on which to base this filtering. What you need is to think about your main or controlling idea: what is the main message you wish to convey? Once you have determined what your main idea is, you will need to group ideas according to this idea, pick the good and relevant ideas, and discard those which, at least for the time being, seem unnecessary, tangential or just irrelevant. This pruning and weeding out of ideas is necessary but also tentative here, as modifications of a different kind are likely to occur at other stages of the writing process. Once ideas have been grouped and prioritized, the following step in the pre-writing stage consists of arranging ideas and topics in an orderly and coherent way. When arranging ideas, consider the pattern of information organization (See Chapter 3, section 3.4) that best suits the topic and conveys the main message. At that point, you are ready to design the plan of your document, that is, to outline.

What are the benefits of outlining? In the outline you should synthetically arrange main and secondary ideas in the manner and order that best suits your controlling idea and allows you to express it logically and clearly. An outline is like a skeleton, with the main and secondary ideas acting as bones. In this sense, an outline will serve as a guideline to help you see how ideas are related and will also prevent you from getting lost in the writing process. Outlining can also be very helpful if the document is to be written by several people, as partitions are easy to make and to allocate. Finally, outlining can be a time-saver in the sense that omissions or any imbalance can already be identified at this initial stage. Remember that peer review is helpful and that it can also be done at this stage.

After you have reconsidered and rearranged topics, you will end up with an outline reflecting your thoughts and your purpose. Later on, in the writing stage itself, this skeleton will be filled with flesh: sentences will be composed and paragraphs will be created.

There are different types of outlines, but we can roughly classify them into two main groups. The first group could be labelled the traditional outline, and the second the visual outline.

Both work well but different writers have different preferences so you should choose the kind of outline that makes you feel more at ease. In traditional outlines ideas can be synthetically expressed using different forms like a key word, phrase or even a sentence. Depending on the variety of the traditional outline, Roman numerals, decimals or letters can be used to show the sequence in which the ideas appear; indentations reflect which ideas are equal and which are subordinate or secondary. See Figure 2.3 below for an example of a traditional outline.

Main Idea: The two sides of the Internet

I. Introduction

II. Positive facts...*first main idea*

    A. Many useful applications...............................*subargument or secondary idea*

        A.1. gateway to information.....................*first proof of secondary idea*

        A.2. variety of communication resources

        A.3. access to research communities

        A.4. new channel for business

    B. Advantages of the Internet.............................*subargument or secondary idea*

        B.1. quantity of information

        B.2. accessibility

        B.3. affordability

        B.4. immediacy

III. Negative or unsolved facts..*second main idea*

    A. Bad use or abuse of the Internet..............................*subargument*

        A.1. hackers

        A.2. pornography

        A.3. illegal trade and fraud

        A.4. addiction

    B. Disadvantages of the Internet ....*subargument*

        B.1. constant change

        B.2. no organization

        B.3. poor /fake information

    C. The Internet in developed and underdeveloped countries ....*subargument*

        C.1. accessibility restricted to developed countries

        C.2. need to develop technological infrastructure
            in underdeveloped countries

        C.3. no real globalization, no real worldwide communication

IV. Pros and cons of a possible regulation or supranational surveillance entity.... *third main idea*

    A. Arguments in favour.......................................*subargument*

        A.1. control of illegal transactions and fraud

        A.2. protection against hackers and other abuses

        A.3. increase of privacy and security

    B. Arguments against..*subargument*

        B.1. who will control the Net?

        B.2. who will set the criteria for censoring information?

        B.3. would this hinder freedom of expression?

V. Conclusion

Fig. 2.3 Traditional outline

The main benefit of **visual outlines** is that, as the name indicates, information is displayed more visually and more synthetically than in traditional outlines. Among the most typical varieties of visual outlines are the *nucleus outline*, or the *tree node*. In the **nucleus outline**, main ideas are circled and secondary or supporting ideas derived from it are indicated by means of arrows, and in the **tree node**, a diagram is used with straight lines which act as branches and sub-branches. Figure 2.4 below is an example of a visual outline.

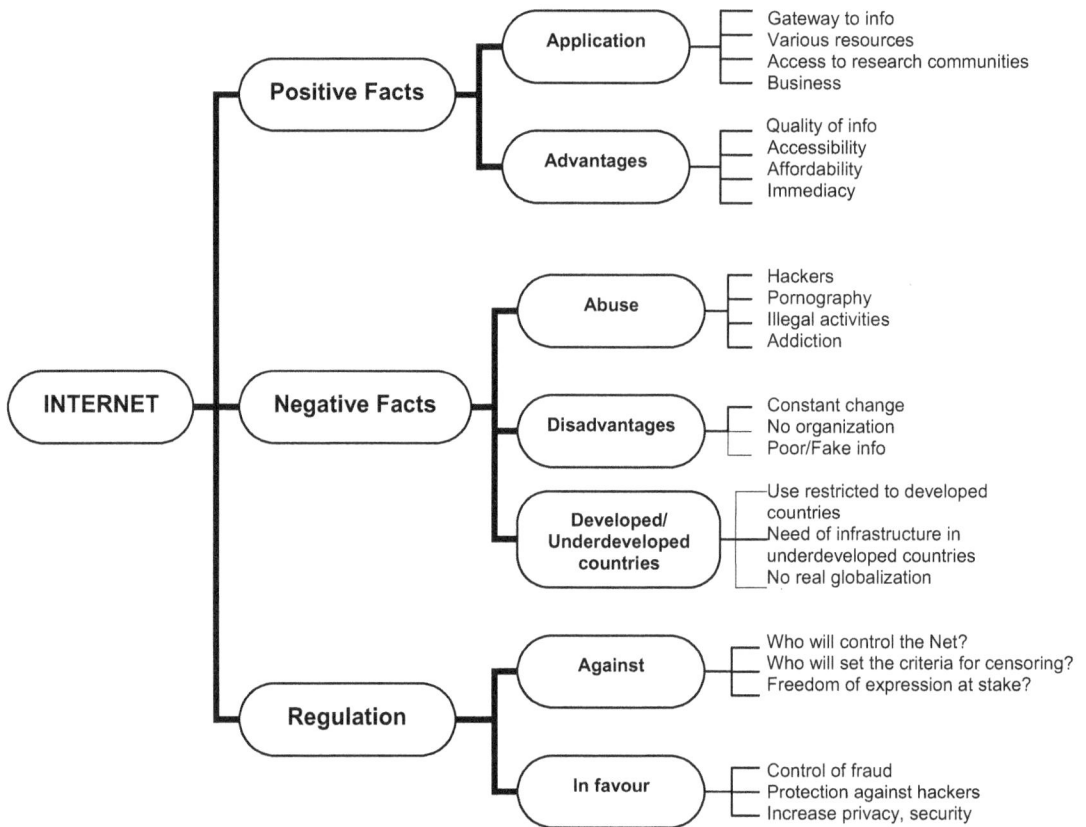

Fig. 2.4 Visual outline

There are two more points to remember when outlining. First, different sections in an outline should be balanced in terms of length; that is, a similar number of ideas should appear in every section. An imbalanced outline will probably lead to either underdeveloped paragraphs or excessively long ones, so it is worth redressing an imbalance at this stage than later when drafting. Second, you should also be coherent in the grammatical form you choose; for example, if you decide to use a gerund form when expressing a secondary idea, then be consistent and use this form with all the remaining secondary ideas.

INDIVIDUAL TASKS

2-32 You have been told to write an essay on the role of *Alternative sources of energy nowadays*. With the ideas generated, try to set up an outline. Choose the type of outline you prefer.

2-33 Imagine you are asked to write a report on the academic and personal progress you have made since you entered your school or faculty. The main idea you want to put forward in the outline is your evolution or development following a chronological order.

2-34 Imagine a friend asks you to review his/her outline. Briefly scan it and decide:
- whether it could be improved or not. If so, indicate in which way or ways.
- how the topic could be approached in a more neutral way without the writer taking sides. Sketch another outline that reflects this other approach to the same topic.

Main idea: pros and cons of mobile phones

I. Introduction
II. Pros of mobile phones
    A. immediacy
    B. ease and comfort
III. Cons of mobile phones:
    C. need to be recharged
    D. more expensive than traditional phones
    E. loss of privacy
    F. exposure to microwaves: risk?
IV. Conclusion: they have become an indispensable gadget

COLLABORATIVE TASKS

2-35 Re-read the text *'Stuff you don't learn in engineering school'* in the previous chapter and in pairs sketch a traditional outline. Remember to keep the same grammatical category within the same group or subgroup of ideas.

2-36 Study the table below on the development of the world's population growth. Then in pairs provide two different outlines on the same topic, one according to a problem-solution approach and another according to a cause-and-effect approach.

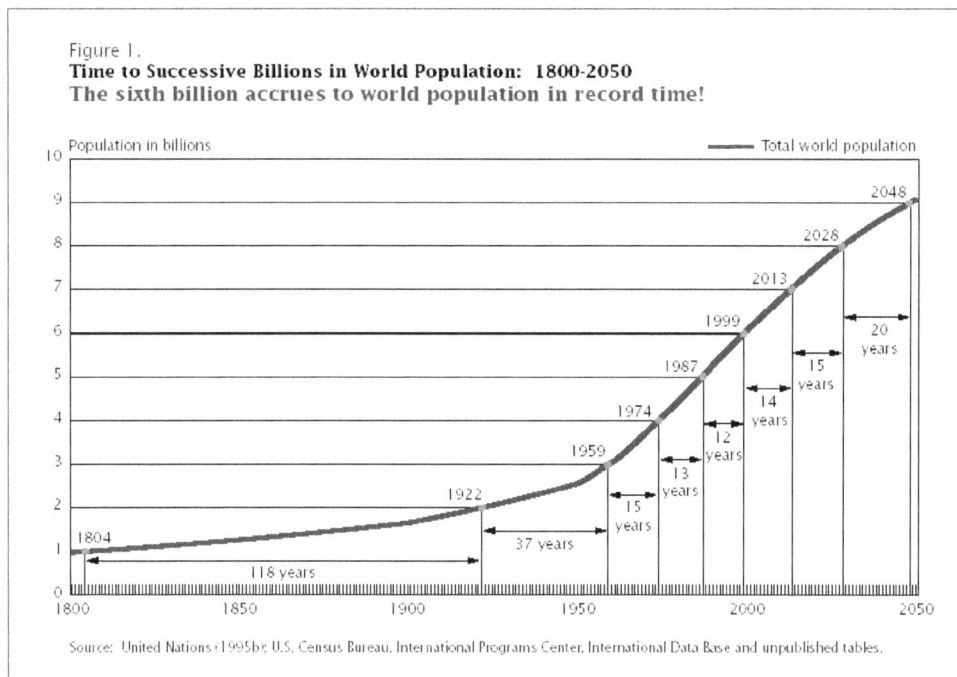

Figure 1.
Time to Successive Billions in World Population: 1800-2050
The sixth billion accrues to world population in record time!

Population in billions ———— Total world population

(graph showing world population growth from 1800 to 2050)

- 1804 — 118 years
- 1922 — 37 years
- 1959 — 15 years
- 1974 — 13 years
- 1987 — 12 years
- 1999 — 14 years
- 2013 — 15 years
- 2028 — 20 years
- 2048

Source: United Nations (1995b); U.S. Census Bureau, International Programs Center, International Data Base and unpublished tables.

PROJECT

This project is developed through chapters two, three and four to help you work through all the different steps of the writing process described in this book. As we have already mentioned before, the writing process developed here is not intended to be prescriptive. However, it may be a good idea that you cover all the different steps and apply the suggested strategies and techniques in order to choose those that best suit your needs and learning style. At the end of these three chapters you will find a description of the specific tasks to carry out at each stage of the writing process. These tasks will be approached from two different perspectives: bottom-up or top-down. You may choose any of these two approaches or you may decide to develop both of them in order to see which is more suitable to your writing style. With the bottom-up approach, you will begin by analyzing the context in which your document will be read and you won't begin writing a first draft until you get to the writing stage. If you choose the top-down approach, you will begin by writing a first draft and then you will revise and rewrite your text, applying the techniques covered, in order to improve that first version. All the different tasks will be collected in a portfolio so that you can see the development of your work.

STEP 1

Let's begin with the pre-writing stage. In this first step you have to choose a topic of your interest related to your field of studies. This is the topic you will write about in order to develop your project, so think about it carefully in order to make a good choice. This first step is common to both approaches, but from now onwards you will find different instructions depending on the type of approach chosen.

Bottom-up approach

STEP 2

After choosing a topic you should continue with the remaining steps of the pre-writing stage. Here you should:

- *Define the profile of the audience of your intended document.* You can do that with the help of the checklist on page 50. This will help you analyze and obtain useful information in order to successfully target your document. Then determine the *purpose* of your document by using the questions on page 51.

- *Decide the most appropriate style and tone.* In order to make a decision remember to take into account the audience's needs and interests.

- *Gather information and ideas for your document.* This is an important pre-writing step as it will determine the contents and also the quality of your text. Therefore, at this stage, you should research different sources in order to collect information. Always remember to cite the sources your arguments are based on. In addition, you can also try some of the above techniques used to generate ideas. These two different ways of collecting information and ideas will help you overcome blank page syndrome and get started.

- *Arrange all these ideas in an orderly and coherent way.* At this point you are ready to *outline*; that is, you should design the plan of your document in the way that best suits your topic. Hence, you should arrange main and secondary ideas so that they serve as a guideline and prevent you from getting lost in the writing process. As you know, there are different types of outlines, which can be roughly classified into traditional and visual outlines. Since all of them work well, choose the one that best suits your preferences and that makes you feel more at ease.

Top-down approach

STEP 2

In this approach, after deciding on the topic, you should write a first draft of your document. Then:

- *Make sure it accommodates the intended audience and purpose.* You can use the checklist on page 50 and the questions on page 51. If you detect any deficiencies, correct them to adapt the document to your chosen audience and purpose.

- *Revise the tone and style you have adopted.* See if they are appropriate for your audience. Again, if you detect any mismatch, rewrite the parts of your document where the style or tone adopted is not suitable.

- *Validate the information and ideas in your text.* You should check if you've written with authority on the topic and given enough support to your arguments. If you need to include more details or quote other authors, remember to cite the sources. In addition, check if your document includes all the information it should include, otherwise it will be ineffective and inadequate. Once you've revised these aspects rewrite any parts of your text with deficiencies. Also remember not to abandon this idea-generating process at later stages as it may always help you further develop some points.

- *Identify any omissions or imbalance in your document.* Developing an *outline* will help you see how your ideas are related and how your message is conveyed. Therefore, write an outline from your text synthetically arranging main and secondary ideas. This will help you spot any gaps and unevenness in the way you develop your controlling idea. Finally, rewrite any part of your text which does not express your message logically and clearly.

CHAPTER 3

Writing stage

Reflecting on...

Whenever you have written a text in English, what do you prefer, writing a first and unique draft or writing several rough copies? List the advantages and disadvantages of each system.

How would you improve a text that has been written in a telegraphic and choppy way?

In your opinion, do you think there's any relationship between what you write and how you write it? Which of these two aspects is more important? Do you think that a document with good and innovative contents but poorly and incoherently written will receive the same credit as if it were clearly and accurately written?

In order to write a text in a clear and readable way, what sort of techniques would you use?

3.1 Introduction

Once you have analyzed audience and purpose, gathered data and outlined the main ideas, you are ready to write the full text. However, what you are actually going to write is not a final, perfect version of your document but a first or rough draft. A version in which what matters is that the ideas sketched out in the outline are fully and clearly expressed in order for the reader to easily understand them. So, at this stage you needn't be over concerned about grammatical or organizational mistakes that may slip into your writing; the important point here is to put your ideas across as effectively as possible. Grammatical accuracy, editing and proofreading aspects must be left for later versions when revising the document. Given that the writing stage is mainly concerned with developing ideas into paragraphs and texts, it becomes necessary to focus on: a) structural considerations, b) organizational considerations and c) linguistic aspects like coherence. It should be pointed out that the approach chosen to present these features begins with the paragraph, which is the smaller unit, and ends with the essay, which is a larger unit. We believe this is a practical and didactic way of organizing the information that follows.

3.2 Drafting

We could say that we've come to the nitty-gritty of the writing process; that is, the stage where the actual writing begins. At this point, the collected information, which was schematically ordered in topics and subtopics, will be fully developed into sentences and paragraphs. In other words, the ideas sketched out in the outline will be expanded into complete sentences, which, in turn, will be organized into well-constructed paragraphs following any of the patterns of information organization[5] that the writer thinks best suits the content. For instance, the widely-used patterns of problem-solution or cause-effect have been recognized as suiting very well the content of technical texts. So choosing the most appropriate pattern of organization is an important factor that will contribute to clarity and readability. You should remember that one major point to bear in mind when writing is that the information should be organized in a clear and logical way so that the reader can readily follow the writer's train of thought. You cannot assume that the reader knows what the writer knows. Besides, technical texts, in general, develop complex and sophisticated ideas the logic of which cannot be easily understood by the reader unless it is logically presented. In this logical way of developing ideas, the writer can be compared to a driver leading the reader along a straight and pleasant road free of obstacles (unnecessary information) and misleading crossroads (confusing points).
Drafting entails accounting for different aspects of the writing process, mainly structural and organizational, among which are:

---

[5] Patterns of paragraph development will be dealt with in Section 3.4

- the format we want to adopt: how can we make the text more understandable and plain? Can we insert visual aids?
- the level of specificity we want to reach: do we want to include examples as well as other clarifying techniques such as definitions?
- the organizational pattern we want to use: do we want all the paragraphs to follow the same pattern?

Another way to keep the logic within and between paragraphs and to make them more readable is by using connecting expressions that explicitly signal semantic relationships. It is not enough to place one sentence after another all the time; there must be some sort of linker or explicit indicator showing the relationship between sentences and paragraphs. Paragraphs and texts need what is known as *coherence* and one major way of achieving it is by using linking expressions. Transitions between ideas must be provided if we want to prevent the text from sounding telegraphic and choppy.

It is also important to bear in mind that, especially during the first draft, all the different aspects involved in this stage (those aspects related to structure, organization and coherence) should be considered together. None of them comes first, none is most important, but rather they are considered synchronously. Not until later, during the rewriting stage, will we proceed to check for the different aspects separately, one by one.

Finally, remember that nobody writes a perfect first draft. On a second reading we all feel that certain parts should be rewritten, others should be taken out and others should be added or just expanded. So don't get stuck with your first draft trying to write a final version of the document. Remember that a draft should be just a draft, no more than this.

3.3 Structuring the paragraph

Now it is time to begin expanding the ideas you gathered and sketched in the pre-writing stage. These ideas should be developed into full form; that is, into a paragraph. A paragraph is a unit of writing constituted by a group of sentences that together express a main idea. A paragraph can stand by itself but it is normally found as part of a larger unit: the text. In fact, it is the visual element which breaks up the text into manageable units of information. Its length is arbitrary: it depends on how much support the main idea of the paragraph needs, so it can range from few to many lines. However, for a question of effectiveness it is advisable to write paragraphs that are neither too long nor too short in relation to the length of the text. For instance, a paragraph that develops a main idea from which some complex and dense subideas stem is likely to be too long, so you should consider dividing it into shorter paragraphs devoting one paragraph to each subidea. Also remember that a paragraph is a visual aid to the reader and that excessively long paragraphs are not visually appealing since they are more difficult to follow. On the other hand, a paragraph containing only two or three sentences is bound to be underdeveloped and calls for more elaboration.

A well-constructed paragraph must display three main characteristics. The first is **development**. Developing a paragraph implies including the necessary information to make the main idea fully understood. From this, it is quite evident that one or two sentences will not be enough. A complete development will generally expand along several sentences that will include enough details to satisfy the questions raised by the topic sentence (see below).

A second important aspect is the paragraph **pattern of organization**. This implies that the information included in the development usually follows a specific pattern of organization. Different topics will be better tackled using different patterns common in technical discourse (cause-effect, problem-solution, general-to-specific, etc). Hence, it is important to select the most appropriate pattern of organization to help the reader better understand the paragraph.

The third characteristic is **coherence**. Coherence in a paragraph is achieved by means of thematic unity and linguistic items (such as for example repetition of key words, use of pronouns and synonyms, linking expressions, etc.). Thematic unity means that all the sentences in the paragraph must be related to the main idea or topic. The introduction of other major points not related to the topic of the paragraph will certainly distract the readers' attention and may leave them confused. Besides, transitions between ideas sometimes need to be explicitly expressed by means of linking expressions or other linguistic indicators since readers cannot always infer the semantic relationships established by the writer; they need to be told how to move from one thought to the next. Such linguistic items, therefore, also contribute to giving coherence and readability to the paragraph by creating a logical flow of sentences.

Structure of the paragraph

Just as a text has a certain internal organization, so does a paragraph. An effective paragraph must display a structure consisting of a **topic sentence**, normally opening the paragraph, some subsequent **supporting sentences**, and finally the **concluding sentence**, which is more optional.

The **topic sentence** introduces the main idea of the paragraph; it is the sentence that tells the reader what the paragraph is about. It is normally the first sentence of the paragraph though, less frequently, it may also be found at the end. The topic sentence is more general than the sentences succeeding it and contains the topic of the paragraph as well as the key words (or controlling ideas) that will be developed in the following sentences. Although it is a good idea to introduce the paragraph with a topic sentence, sometimes the topic sentence is implied or even missing.

The **supporting sentences** are those that develop the topic sentence and constitute *the body* of the paragraph. They can develop the topic sentence in different ways: by defining the key terms, by giving examples or reasons, by introducing facts, etc. Their purpose is to make the topic sentence clear and convincing.

The **concluding sentence** is commonly found in isolated paragraphs in order to indicate that the paragraph has reached the end. They are also used to restate the main idea or to emphasize an important point. However, when a paragraph is found within a text, these sentences are not so necessary and, therefore, are scarcer. In some instances, they act more as a transition between paragraphs than as a real or firm conclusion to the topic.

The figure below shows two possible ways of organizing information in a paragraph.

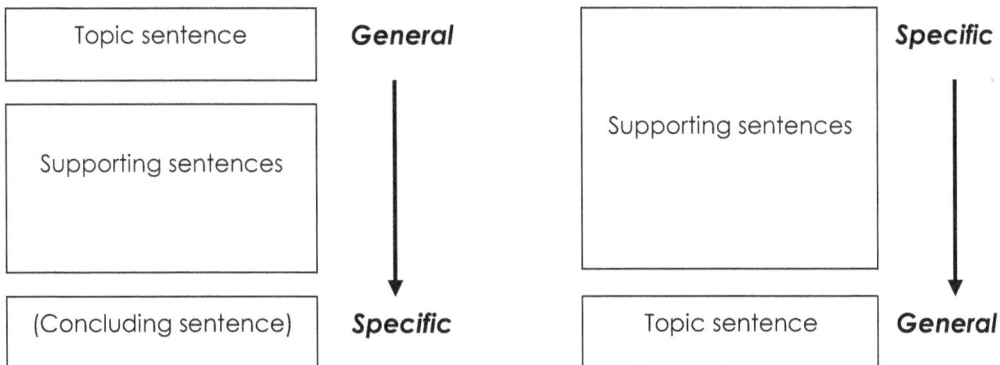

Fig. 3.1 The paragraph structure

The following is a good example of a paragraph with the topic sentence at the beginning followed by the corresponding supporting sentences and a concluding sentence at the end.

Topic sentence	**Digital systems have substituted analog systems in many new personal technologies**. Although analog systems are more reliable and accurate than digital ones, the fact that digital systems are cheaper, smaller and faster makes them very suitable for many personal technologies. Thus, while analog
Supporting sentences	systems are more adequate to technologies that require extreme accuracy as for instance precision-sensor systems, digital systems fit better those technologies that need speed and easy-to-use information such as computers. Computers are a common example of personal technology that has allowed to process huge amounts of information in fractions of time thanks to the digital
Concluding sentence	system they use. **So the advantages of digital systems and their multiple applications are what make them so valuable.**

The topic sentence

When writing a topic sentence there are some important points to remember:

- *A topic sentence must be a complete sentence*; that is, it must have a subject, a verb and (complements). The use of expressions without a verb is inadequate and grammatically incorrect. In fact, they are closer to titles than to topic sentences. The following expression exemplifies this:

    The importance of technological advances in modern life. (*Incorrect*)
    Technological advances are very important in modern life. (*Correct*)

- *A topic sentence must have two main parts: the topic and the topic development*. The *topic* introduces the theme or subject of the paragraph. The *topic development* indicates what we want to say about the topic. It restricts the topic by specifying the aspects that will be dealt with in the paragraph.

> <u>The heavy traffic</u> is becoming <u>a serious problem in big cities</u>.
>   TOPIC   TOPIC DEVELOPMENT

The topic is normally placed at the beginning of the sentence, but can also be found somewhere else. Here are three examples with the topic in different positions:

> *Example 1*: **E-mailing** has several advantages over other communication systems.
> *Example 2*: There were much slower **systems of processing information** before than today.
> *Example 3:* Many people have a wrong idea about the **use of microwaves.**

- *A topic sentence must not be too specific.* If it is too specific, there will not be any information left to be included in the rest of the paragraph, so the paragraph will end just there, after the topic sentence. For example:

> Solar cars are not very practical because they need to recharge their batteries very often and because they need sunlight. (*Incorrect*)
> Solar cars are not very practical for two main reasons. (*Correct*)
> Solar cars are not very practical because of their batteries and power source. (*Correct*)

However, as reflected in the two correct topic sentences above, there is a certain degree of freedom in relation to the specificity of a topic sentence. In the first sentence, the writer has chosen to be more general whereas in the second he/she has preferred to be more specific and outline the two main reasons why solar cars are not very practical.

- *The topic sentence can be found at the beginning or at the end of the paragraph.* Nevertheless, it is more common to find it at the beginning because this is a way to help and better orient the reader about the content of the paragraph, and also because it is a way to link the paragraph with the previous one in a text. The following is an example of a paragraph where the topic sentence is at the end:

> On March 16, 1978, The AMOCO CADIZ ran aground off the coast of Brittany, France, spilling 68.7 million gallons of oil. On March 24, 1989, the oil tanker EXXON VALDEZ struck Bligh Reef in Prince William Sound, Alaska, spilling more than 11 million gallons of crude oil. On November 19, 2002, another oil tanker, the PRESTIGE, sank about 135 miles off the coast of northern Spain and Portugal taking down with it at least 18 million gallons of fuel. Before going down, it spilled almost two million gallons of heavy fuel oil. **These are some examples of major oil spills caused by human error, which have brought about important environmental pollution**.

- *The topic sentence can indicate how the paragraph will develop by advancing the pattern of organization that will follow.* An example of this is:

> Digital systems process information faster than analog systems.

This topic sentence suggests that the pattern developed in the rest of the paragraph will be based on comparison and contrast by specifying the ways the two systems process information.

Note that a useful way to check whether a sentence is a good topic sentence is by trying to ask possible questions, the answers of which will be the specific details that will be included in the supporting sentences. So for a topic sentence to be a good topic sentence, it must allow you to ask questions: How?, Why?, When?, What?, In which ways?, etc. For example, the following sentence

Friction may be destructive and wasteful.

allows you to ask: Why may friction be destructive and wasteful? The reasons you will give to support this idea will constitute the body of the paragraph. But if you write

Friction may be destructive and wasteful because it may damage components and cause breakdowns.

and you try to ask a question, you will realize that there is no possible question to be asked since everything has been said in this sentence. It is too specific to be a topic sentence.

Supporting sentences

As pointed out before, a well-constructed paragraph has a topic sentence and some supporting sentences that develop the main idea introduced in the topic sentence. But how can the supporting sentences be arranged? Is there any specific way to arrange them? Indeed, there are two main ways of organizing them. In the first pattern all supporting sentences are directly related to the topic sentence.

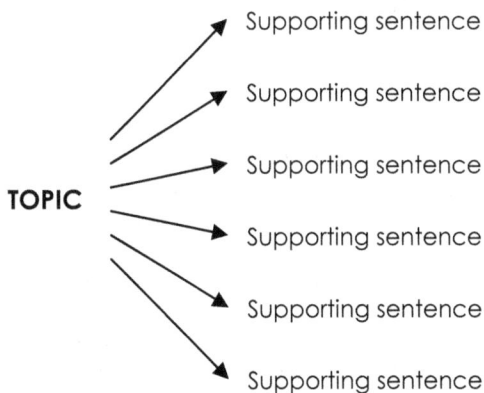

In the second pattern not all supporting sentences are directly related to the topic sentence. Some of them derive from the previous supporting sentence. This is the case of those sentences that need examples, details or some kind of specification.

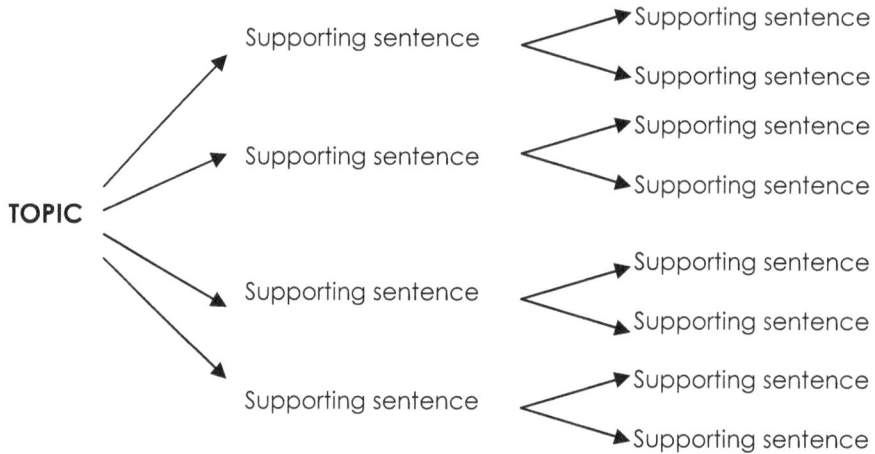

These two different patterns will be visually represented in a paragraph format, in one block. However, in technical texts it is also quite common to represent these ideas using a bulleted list.

INDIVIDUAL TASKS

3-1 Read the following group of sentences and divide them into three paragraphs. After identifying the three paragraphs, provide a topic for each. Finally, can you identify any sentence acting as a transition between paragraphs?

Digital Versatile Disc

Since the Audio compact Disk and CD-ROM were introduced in 1982 and 1985 respectively, compact discs (CDs) have become popular media formats, and are used to carry music, data, and multimedia entertainment (AME). When the CD-ROM was developed, it had the ability to store over 650 megabytes (MB) worth of data or music (Disc Manufacturing, Inc.). Today, however, the capacity of 650 MB of storage is too limited for computer applications. As a consequence, a second-generation disc technology is needed to provide video, multimedia, and databases more quickly and in greater volume. In 1995, the successor to CD, DVD, was announced. Every DVD disc is made of two parts, each of which is 0.6 mm thick; thus, together, the two parts are 1.2 mm thick, which is the thickness of a current CD. Compared to a CD, which has the ability to store 650 MB, DVD holds seven times the data. The store capacity of a Single Sided/ Single Layer DVE is 4.7 gigabytes (GB) "That's enough room to store 133 minutes of full-motion video per side" (Normile56). Single Sided/ Dual Layer, which is expected to be the most popular configuration, can hold a special edition

| DVD video movie or 8.5 GB of computer data (Disc Manufacturing Inc.). Another configuration is Double Sided/ Single Layer, which can hold 9.4 GB of computer data or one movie plus 4.7 GB of data. The last one is Double Sided/ Dual Layer, which can store 17 GB. This disc could pack up to four movies (Normile 57). Owing to the data capacity, DVD will be able to provide multiple language and subtitle tracks, which allow users to choose whether to listen to the original movie dialogue, with or without subtitles, or to a dubbed version (Normile 57-58). Moreover, in theory, the viewer would be able to control how a scene is viewed, choosing from as many as eight different camera angles (Vizard 71).

Source: Kritsilpe, Y. (n.d.). Digital versátile disk. [WWW page].
URL http://www.tcomschool.ohiou.edu/its_pgs/dvd.html

3-2 Bearing in mind the distinction between the topic and the topic development, identify the topic of the following topic sentences.

1. The future use of hydrogen cars is mostly subject to political and economical reasons.
2. It is very difficult to protect computers from incoming viruses.
3. Practising sports is necessary for different reasons.
4. Modern lifestyle has brought many advantages but also many disadvantages.
5. Driving too fast can carry serious consequences
6. The use of laser has found many applications in medicine
7. Nowadays we already have some evidence that the high emission of CO_2 is affecting the Earth.
8. Acceptance for publication depends mainly on the significance of the study (content) and the written form.
9. There are several effective ways to transmit information using electric signals.
10. The invention of the chip was one of the most important technological breakthroughs.

3-3 Indicate whether the following topic sentences are correct or incorrect and justify your choice.

1. I think that water reserves are becoming almost depleted in some parts of the world.
2. This study demonstrates the advantages and disadvantages of using solar energy.
3. Biotechnology is one major field of study because it can make important contributions to the medical community.
4. The impressive Three Gorges Dam is located on the Yangtze River in China.
5. Controlling TV programs is good for children as well as for adults.
6. The main differences between the analogical and digital communication systems can be found in an introductory Communication Systems book.
7. Silicon was found to be very useful in developing chips because of its high resistance and low cost.
8. Sending information by email is much cheaper than telephoning.
9. Comparison between optical fibre and coaxial cable as regards the speed of transition and cost.
10. Writing is a hard task but it is also very satisfying.

3-4 Find the sentence that is irrelevant to the topic of the paragraph and delete it.

> Volcanoes can be classified into three main groups: active, dormant and extinct. The traditional criterion for this classification was the date of the last eruption. If the last eruption fell within historic times– the period people have been recording history—the volcano was deemed active. If the last eruption occurred before historic times but within 10,000 years, the volcano was considered "dormant", because it likely had the potential to erupt again. Volcanoes that had not erupted in more than 10,000 years were considered extinct, because it seemed unlikely they would erupt again. However, volcanoes are a crucial element of the earth's ongoing regeneration.

> Source: Harris, T. (January 15, 2001). How volcanoes work. [WWW page]. URL http://science.howstuffworks.com/volcano2.htm. (February 03,2007). Courtesy of HowStuffWorks.com.

3-5 Choose the topic sentence you think best fits each of the following paragraphs

Sample A
First generation computers appeared just on the onset of the Second World War. They were characterized by the fact that operating instructions were made-to-order for the specific task for which the computer had to be used. Each computer had a different binary-coded program called "machine language" that told it how to operate. This made the computer difficult to program and limited its versatility and speed. Another characteristic of the first generation computers was the use of vacuum tube magnetic drums for data storage.

a)   First generation computers were very useful.
b)   First generation computers were slow and difficult to program.
c)   First generation computers had distinctive features.

Sample B
Every year hundreds of people die in road accidents. Although it is compulsory to wear seat belts, some people either forget to use them or overlook the law because they feel uncomfortable with them. Also, despite the striking TV campaigns, people are still little law-abiding with regard to alcohol consumption and sometimes even drive after drinking a considerable quantity of alcohol. Another important cause of road accidents is speed, even though with the current technology it would be easy to detect cars exceeding the speed limit. So the courts should enforce adequately current legislation and the police should ensure that the laws are observed.

a)   Driving entails many risks.
b)   There are many factors that contribute to the problem of road safety.
c)   Some people like overlooking certain traffic laws, namely drinking alcohol and exceeding the speed limit.

Sample C
Well-constructed nuclear power plants have an important advantage when it comes to electrical power generation—they are extremely clean. Compared with a coal-fired power plant, nuclear power plants are a dream to come true from an environmental standpoint. A coal-fired power plant actually releases more radioactivity into the atmosphere than a properly functioning nuclear power plant. A coal-fired plant also releases tons of carbon, sulphur and other elements into the atmosphere. Another important point to take into account is the fact that coal-fired plants need large amounts of coal to generate electricity because only 40 percent of the thermal energy in coal is converted into electricity.

Source: Brain, M. (April 01, 2000). How nuclear power works. [WWW page]. URL. http://science.howstuffworks.com/nuclear-power1.htm. Courtesy of HowStuffWorks.com.

a) Nuclear-power plants are quite different from coal-fired plants.
b) Nuclear power plants are more beneficial than coal-fired plants with regard to environmental pollution.
c) Nuclear-power plants are easier to clean than coal-fired plants

3-6 These are some topic sentences from which you can develop complete paragraphs. Choose a topic sentence of your interest and write a paragraph. Consider the key words in the different sentences as they will help you include the necessary details to make the main idea clear.

- Automobile safety is becoming one of the major concerns of auto manufacturers today.
- An important application of satellites is found in communication systems.
- In comparison to LPs, CDs have a better sound recording quality.
- Computers have undergone an incredible evolution since the first models appeared.
- Nuclear power has had not only military applications but peaceful applications as well.
- Latex has many industrial uses.
- There is more acid rain today than at any other time in history

COLLABORATIVE TASKS

3-7 Read the paragraphs and then in pairs write a good topic sentence for each.

Sample A
The main reason for the doubling of the world's population during the second half of the 20th C. is a result not so much of a rise in births but of a fall in deaths. Thanks to the improvements of the public health services and medical care, people of all ages live longer. Many more babies now outlive infancy and grow up and many more adults reach the old age so that population increases at both extremes. But the death rate

began to fall differently in different parts of the world. While in Europe and North America it began to fall during the Industrial Revolution, in Africa, Asia and Latin America the fall in the death rate did not begin until much later.

Sample B

Computers are being used more and more extensively in the world today because they are more efficient than human beings. Their incredible speed makes it possible to solve complex problems in a small fraction of time. For example, they are used in business to keep accounts and in airlines, trainlines and buslines to keep track of ticket sale. Additionally, their memory capacity allows them to store huge amounts of information with high accuracy and reliability. For example, a computer can store as many as a million items and can put information into a computer and retrieve it in a millionth of a second.

Sample C

First of all, the installations and equipment necessary for most alternative sources of energy are very expensive. This is the case with nuclear power stations that, despite needing very little fuel to produce enormous amounts of power, are very expensive to build and the governments are unwilling to invest in them. Secondly, alternative sources of energy are not as reliable as the traditional ones since the storage of energy depends on uncontrollable factors such as the wind and the sun. Last but not least, alternative sources of energy need highly-qualified workers in order to run the system properly.

3-8 Read the following scrambled sentences. In pairs, identify the topic sentence of each paragraph and order and combine the rest. Use linking expressions where appropriate in order to get a good paragraph.

Paragraph 1

The *human error* is found in a very high proportion of maritime causalities and often neither hardware nor the most advanced technology can eliminate it.

We should work hard to reduce the risks associated with maritime disasters.

The rapidity and effectiveness of marine communications may be affected by several factors: hardware, lack of discipline in using standardized procedure, and language barriers.

The losses in terms of lives and properties and the damage to the environment cannot be afforded. The first factor depends on technology and the last two depend on human beings.

Whereas the first factor depends on technology, the last two depend on human beings.

Paragraph 2

Weather forecast satellites use a rich variety of observations from which to analyze the current weather patterns.

The launch of the first weather satellite in 1960 made possible global observations, even in the remotest areas.

Satellites are essential in weather forecast.

Nowadays, it is possible to make a short-term weather forecast (5 days) and even a long-term forecast (3 months).

These forecasts are extremely useful to predict cyclones, big storms and other catastrophes.

Paragraph 3

The acoustic noise originates from the engine, air flow, frictions and vibrations.

The acoustic noise is coupled to the transmitted speech signal through the microphone, and therefore deteriorates the signal to noise ratio at the transmitter.

Background acoustic noise is a major impairment in voice mobile communications.

This problem is particularly significant in hands-free systems where the primary microphone is located far from the speaker.

The effect of background noise is even more serious in integrated voice-data systems where the performance depends critically on the operation of speech detectors in a noise environment.

CRITICAL THINKING

3-9 Do you think that each and every paragraph in a text must have a topic sentence? What would be the effect of a text with these characteristics?

3.4 Developing paragraph patterns

In the previous sections you learned how to write a well-constructed paragraph, beginning with a suitable topic sentence and following with a number of logically-linked supporting sentences. In this section we'll see that these supporting sentences should also be organized using a consistent pattern that arises naturally from the topic sentence and that helps the reader process the information in the paragraph more easily. This way the writer will choose the pattern that best suits his/her writing purpose and type of document and so the reader will be able to follow effortlessly the writer's train of thought.

Given that technical texts tackle scientific or technical topics, it stands to reason that in technical texts scientific reasoning is reflected in the way information is organized. A general way of establishing a common ground with readers in technical communication is to base arguments on a rational appeal; that is, to base arguments on logical reasoning from evidence. As opposed to texts with an emotional or ethical appeal, technical texts have been demonstrated to follow certain patterns of organization that better capture the scientific way of thinking. Some of the most common patterns of paragraph development used in technical and scientific English are *chronological or sequential order, cause and effect, comparison and contrast, analogy, specific to general, general to specific, problem and solution, listing and order of importance.* All these different paragraph structures may aid to organize the information you are planning to develop in an orderly manner. Also, each pattern usually displays certain specific features which may ease the task of writing and at the same time help the reader identify the pattern used.

Chronological or sequential order

In a chronological description the different events are organized with respect to their occurrence in time. This is a well-known pattern of organization in scientific and technical writing as it is often used to describe processes and procedures (e.g. experiments, instructions, problem-solving procedures, test procedures, etc.) and also to narrate a sequence of past events (e.g. in a progress report or in the review section of a research article). In order to write a chronological or sequential paragraph, first plan what to write, then sequence your events, and finally connect them by using any of the following chronological signals and expressions:

Chronological signals and expressions

Sentence connectors	first(ly), second(ly), third(ly), last(ly), next, then, finally, eventually, after that, afterwards, subsequently, in the meantime, meanwhile, simultaneously, at the same time
Subordinators	before, after, when, whenever, until/till, since, once, as, while, as soon as
Phrase linkers	before, after, when/on/in, previous to, prior to, following
Time adverbs	in 1992, from 1985 to 2005, last month, two years ago, in the next few years, soon
Verb tense choice	First, we *installed...* Up to now we *have developed...* In the future we *will attempt to*
Use of parallelisms	*Preheat* the fuel... *Pump* the fuel using high pressure fuel pumps... *Separate* water and impurities...

Read the following chronological sample paragraphs and study the use of the expressions in *italics,* which indicate time order.

> ### Sample I
> The reflection of radio waves by conducting objects has been used for a long time. It *was first noticed* more than *a century ago. As far back as 1903,* the effect *was used* in

Germany to demonstrate detection of ships at sea. Marconi *championed* the same idea in Britain *in 1922*. However, there *was* little official interest and several years *passed* before systematic experiments in radio detecting *began. Early* work *used* continuous wave transmissions, and *relied* upon interference between a transmitted wave and the Doppler-shifted signal received from a moving target. The detection of aircraft *was first accomplished* in the USA *in the 1930s*. [Italics added]

Source: Lynn P. L. *Radar systems*. (1987). London: MacMillan Education Ltd, 1. Reprinted with permission of Palgrave MacMillan.

Sample 2

In the shipbuilding industry, the process of building a ship is likely to vary somewhat depending on the customer involved but it generally includes a number of specific stages: development of owner's requirements, concept design, contract design, contracting, detail design and planning, and construction. The *first stage* in the shipbuilding process is the costumer's formulation of the product requirements. *Once* the owner has identified the need for a new ship and defined operational requirements, a preliminary definition of the basic characteristics of the vessel is provided. Based on the general description of the ship to be built, as determined by the end product of the preliminary design stage, more detailed information is required to permit contracts to be prepared. *Following* completion of the contract design stage, a specific shipyard is chosen to build a vessel. *After* the bidding process is complete and a contract has been signed, detail design, planning and scheduling proceeds. *The final stage* of the shipbuilding process is the actual construction of the vessel. [Italics added]

Adapted from: Storch, R. L., C. P. Hampton, H. M. Bunch & R. C. Moore. (1995). Ship Production. Centreville, Maryland: Cornell Maritime Press, 3-5.

Sample 1 is an example of a chronological paragraph that recounts a sequence of past events. Notice how the topic sentence introduces the theme, which is then developed following a chronological order. As can be seen, the sequence of events is determined by the use of time adverbs (*a century ago, as far back as 1903, in 1922, in the 1930s*) combined with verb tenses in the simple past tense. Sample 2 includes a paragraph that describes a shipbuilding process. In this example, the topic sentence not only introduces the topic but also provides an overview of the different steps that will be dealt with. To develop the process, the author uses mainly sentence connectors and subordinators to mark the time order.

Cause and effect

Another common way of organizing a paragraph is by means of a cause-and-effect pattern. This pattern is useful when you want to describe a phenomenon caused by some situation or circumstance, or a phenomenon that produced some particular effects. Note that usually if the topic sentence introduces a cause, the supporting sentences analyze effects and vice versa. Another important consideration to bear in mind when developing a cause-and-effect pattern is not to become too simplistic or use clichés. Always try to provide the necessary supporting details to write a consistent paragraph. Choose those causes or effects that you consider most important and maintain your focus all through the analysis. Finally, the logical correlation of a cause-and-effect relationship can be marked by using characteristic signals that will point out the causes and effects in your paragraph. Although an overuse of these signals is not recommended, especially when the relationship is very clear or obvious, your readers will appreciate it if you occasionally introduce some of them as they will help identify the different cause-and-effect relationships more easily.

Cause-and-effect signals and expressions

Sentence connectors	therefore, thus, as a result, as a consequence, consequently, that is why, for this reason, hence
Coordinating conjunctions	so, for
Subordinators	since, as, because, for, the reason why, given that
Phrase linkers	because of, due to, owing to, as a result of, as a consequence of, on account of, the reason for
Causative verbs	cause, produce, result in, give rise to, bring about, result from, be caused/produced by, affect, have an effect on
Conditional constructions	when, where, given, if, then

Below you will find two examples of cause-and-effect paragraphs. Again, note the use of the signals and expressions in italics, which mark the different cause-and-effect relationships.

Sample I

The automotive air-pollution problem became apparent in the 1940s in the Los Angeles basin. In 1952, it was demonstrated by Prof. A.J. Haagen-Smit that the smog problem there *resulted from* reactions between oxides of nitrogen and hydrocarbon compounds in the presence of sunlight. In due course it became clear that the automobile was a major contributor to hydrocarbon and oxides of nitrogen emissions, as well as *the prime cause* of high carbon monoxide levels in urban areas. Diesel engines are a significant source of small soot or smoke particles, as well as hydrocarbons and oxides of nitrogen. *As a result of* these developments, emission standards for automobiles were introduced first in California, then nationwide in the United States, starting in the early 1960s. Emission standards in Japan and Europe, and for other engine applications, have followed. Substantial reductions in emissions from spark-ignition and diesel engines have been achieved. Both the use of catalysts in spark-ignition engine exhaust systems for emission control and concern over the toxicity of lead antiknock additives *have resulted in* the reappearance of unleaded gasoline as a major part of the automotive fuels market. Also, the maximum lead content in leaded gasoline has been substantially reduced. The emission-control requirements and these fuel developments *have produced* significant changes in the way internal combustion engines are designed and operated. [Italics added]

Source: Heywood, J. *Internal Combustion Engine Fundamentals.* (1989). Singapore: MacGraw Hill, 5. Reprinted with permission of the McGraw-Hill Companies.

Sample 2
Internal combustion engines are also an important source of noise. There are several *sources of* noise: the exhaust system, the intake system, the fan used for cooling and the engine block surface. The noise may be *generated by* aerodynamic effects or *may be due to* forces that *result from* mechanical excitation by rotating or reciprocating engine components. Vehicle noise legislation to reduce emissions to the environment was first introduced in the early 1970s. [Italics added]

Source: Heywood, J. *Internal Combustion Engine Fundamentals*. (1989). Singapore: MacGraw Hill, 5. Reprinted with permission of the McGraw-Hill Companies.

These two different paragraphs develop different types of cause-and-effect relationships. Sample 1 begins with the identification of the causes of air pollution in Los Angeles and then goes on analyzing the different effects resulting from this problem. Although this sample basically follows a cause-and-effect pattern, it could also be considered to develop a problem-solution pattern as the effects also constitute the solutions to the problem defined at the beginning of the paragraph. Also, notice how, apart from using different cause-and-effect signals and expressions like *as a result of, have resulted in* and *have produced*, the writer also links different events together without using an overt marker. These different events constitute a chain of actions in which one action leads to the next. This way of narrating events reduces the need to insert too many cause-and-effect signals and expressions.

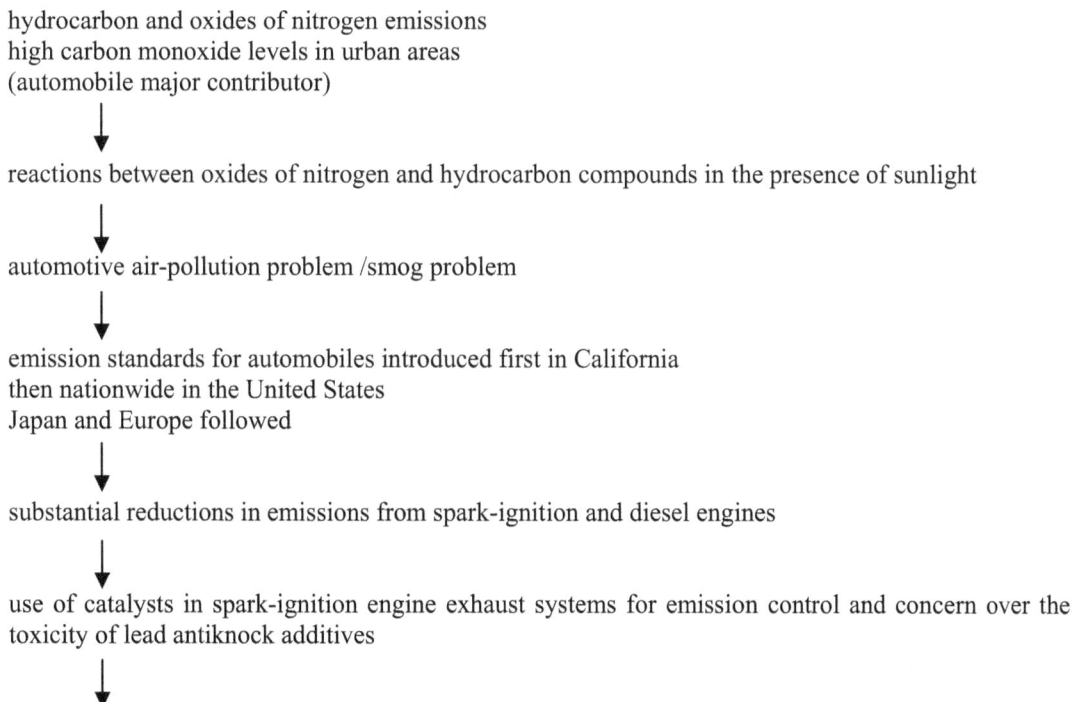

hydrocarbon and oxides of nitrogen emissions
high carbon monoxide levels in urban areas
(automobile major contributor)

↓

reactions between oxides of nitrogen and hydrocarbon compounds in the presence of sunlight

↓

automotive air-pollution problem /smog problem

↓

emission standards for automobiles introduced first in California
then nationwide in the United States
Japan and Europe followed

↓

substantial reductions in emissions from spark-ignition and diesel engines

↓

use of catalysts in spark-ignition engine exhaust systems for emission control and concern over the toxicity of lead antiknock additives

↓

reappearance of unleaded gasoline as a major part of the automotive fuels market
also reduction of the maximum lead content in leaded gasoline

↓

significant changes in the way internal combustion engines are designed and operated

Sample 2, on the other hand, is basically an effect-and-cause paragraph. In this example, the author introduces in the topic sentence an effect produced by internal combustion engines, noise. Once the effect is identified, the causes of this noise are listed and analyzed in the rest of the paragraph.

Comparison and contrast

In a comparison-and-contrast paragraph, two items, methods, etc. are compared in order to show the advantages or benefits of one over the other. Paragraphs developing this sort of pattern usually compare or contrast things that are similar in nature and clearly point out the similarities and/or differences between them. These paragraphs also lay emphasis on the point they try to make by using different comparison-and-contrast signals and expressions. All this helps readers better follow the author's line of argument and clearly identify the winning item resulting from the comparison.

Comparison-and-contrast signals and expressions

Sentence connectors	however, yet, still, nevertheless, nonetheless, on the other hand, on the contrary, by/in contrast, conversely, similarly, likewise
Coordinating conjunctions	but, yet
Subordinators	although, even though, though, while, whereas
Phrase linkers	in spite of, despite, (un)like, in contrast with, as opposed to, different from, compared with, in comparison with, alike, similar to
Correlatives	both ...and, not only... but also, neither...nor
Comparative and superlative constructions	-er than, more than, the –est, the most..., less/fewer than, more than, the same as, as... ...as, not so... ...as
Parallel constructions	The higher the demand, the higher the production Method X is feasible and effective whereas Method Y is more difficult and relatively effective Product Z is not only reliable but also effective and economical

Sample I
The diesel engine comes close to being the universal marine propulsion engine. *Although* it has superior competition for some types of propulsion, it can be used successfully in almost any application, and it is generally acknowledged to be *the best* in a *wider* range of marine applications *than* any other engine. The attribute contributing *the largest share* to the accolade *'best'* is efficiency. Diesel specific fuel consumption in most applications is *lower* (synonymous to *higher efficiency*) *than* *either* steam *or* gas turbine *or* spark ignition (gasoline) plants, and the fuel is usually at

least *as cheap* per unit of heating value *as* theirs. *Compared to* the steam propulsion plant, the diesel also enjoys the advantages of internal combustion, which makes it compact, available in essentially a complete package, and simple to control. The gas turbine and spark-ignition engines are also internal combustion, of course, but the efficiency advantage heavily favours diesel over both of these. The use of a fuel of much *lower volatility than* gasoline also gives it an important safety advantage over spark ignition. [Italics added]

Source: Woodward, J. B. (1981). *Low Speed Marine Diesel.* Malabar, Florida: Robert E. Krieger Publishing Company, 1.

In this example, the author describes the advantages of diesel engines over other types of engine in marine applications. Notice the important number of comparison-and-contrast signals and expressions used to highlight the advantages of this specific type of engine. Besides, the use of some expressions not specific to this pattern (for example *at least, heavily favours, gives it an important safety advantage over*) also contributes to emphasizing the contrast between the different types of engine. That way, the writer clearly states his/her position without leaving any room for doubt in order to persuade the reader.

A pattern that might be confused with a comparison-and-contrast one is that of analogy. Analogies are used to provide readers with a familiar scenario to explain an unfamiliar concept by showing the ways in which they are similar. However, differently from comparison-and-contrast patterns, analogies compare concepts that are basically different in order to familiarize readers with a complex concept, so it may be guessed that they are more typically addressed to a non-specialist audience.

Sample 2
A router is a piece of equipment which takes packets (packages of data), and sends them to where they are trying to go. A simple analogy would be a traffic policeman at a busy intersection. Cars want to go through the intersection, but without the policeman there could be accidents. So the policeman takes control of the intersection, controls when and where the cars can go, and makes sure that no accidents happen. The policeman would be our router, each road which meets at the intersection would be one of the networks connected to the router, and the cars are the packets. Now let's take this analogy a bit further. The policeman spots someone he suspects of being a criminal, a bank robber for example. Fortunately, the policeman is in control of the road leading to the bank! So the policeman then stops the bank robber's car from using the road to the bank, and saves the bank from robbery. In computer science terms, this is known as firewall, and stops hackers from entering networks.

Source: Kakanowski. T. (1997). What is a router and how does it work? [WWW page]. URL http://www.madsci.org/posts/archives/may97/863012468.Cs.r.html.

The sample above is an example of an analogy in which the author describes the working of a router by comparing it to a traffic policeman at a busy intersection. This way, the writer explains in plain and familiar words a rather complex concept for a non-expert audience.

General to specific

A general-to-specific paragraph begins by stating a generality which is then developed and further specified in the subsequent supporting sentences. That is, the topic sentence usually introduces a general overview of the theme that will be dealt with and the supporting sentences provide the details by defining, providing examples, classifying, etc., which may become increasingly more specific. This deductive structure, in which the generalization or core statement precedes the supporting details, is typical of most technical and scientific paragraphs and is often associated with technical descriptions.

> ### Sample I
> Earth materials are often used as a construction material because they are the cheapest possible building material. However, its engineering properties such as strength and compressibility are often naturally poor, and measures must be taken to densify, strengthen, or otherwise stabilize and reinforce soils so that they will perform satisfactorily in service. Highway and railway embankments, airfields, earth and rock dams, leeves and aqueducts are examples of earth structures, and the geotechnical engineer is responsible for their design and construction.
>
> Source: Robert D. & W. D. Kovacs. (1981). *An Introduction to Geotechnical Engineering*. Englewood Cliffs: Prentice Hall, 2. Reprinted with permission of the publisher.

In this example, notice how the first sentence introduces a general statement on the use of earth materials as construction materials. The subsequent sentences develop the topic by discussing their properties and by providing some examples of their applications.

Specific to general

In a specific-to-general paragraph pattern different points are presented and then a general conclusion is drawn from these. This structure reflects inductive reasoning, where facts or observations are assessed in order to make a general statement. Therefore this final general statement may either be a topic sentence or an evaluative statement on applications, advantages and disadvantages, etc. In technical and scientific writing, this pattern is commonly found in executive summaries and proposals.

> ### Sample I
> A piezoelectric effect occurs when certain materials are subjected to mechanical stress. An electrical polarization is set up in the crystal and the faces of the crystal become electrically charged. The polarity of the charges reverses if the compression is changed to tension. Conversely, an electric field applied across the material causes it to contract or expand according to the sign of the electric field. The piezoelectric effect is observed in all ferroelectric crystals and in nonferroelectric crystals that are asymmetric and have one or more polar axes. This effect is important because it couples electrical and mechanical energy and thus has many applications for electromechanical transducers.
>
> Source: Young. E. C. (1979). *Dictionary of Electronics*. Penguin Books, 426. Reprinted with permission of the publisher.

This particular sample begins with a description of the piezoelectric effect. After different facts and observations have been explained, the author ends with a general conclusion which highlights the importance and the large number of applications of this effect.

Problem and solution

In a problem-and-solution pattern, a situation is presented and one or several alternatives may be proposed for resolution. The sequence of information in this structure usually begins with a clear description of the problem, follows with an analysis of the processes or causes contributing to the problem and ends with the identification of the solution. It is particularly important to pay attention to the description of the problem because readers have to be convinced of the existence and /or nature of the problem before solutions are put forward. This pattern reflects a more evaluative and argumentative type of critical thinking that conforms to the reader's expectations about the problem-solving process. Reports and proposals are the types of documents most commonly related to this pattern in technical and scientific writing. Also notice that, as the discussion of the causes of the problem is often included in this sort of pattern, there is a frequent overlap between this structure and the cause-and-effect one. Nevertheless, as will be explained below in the conclusion, overlaps are not exclusive of these two structures, but a constant among the different patterns of paragraph development.

Sample I
Urban air pollution is one of the most worrying environmental problems nowadays. Two factors can be singled out as significant causes of urban air pollution: nitric oxides and hydrocarbons emitted in large part from vehicles on the one hand, and volatile organic carbons (VOC), mainly produced by the volatility of fuel and the manufacture and extensive use of synthetic fertilisers on the other. Fortunately, current technology is trying to reduce these emissions substantially. Emission control strategies involve catalytic oxidation of unburned hydrocarbons, reduction of nitric oxide as well as what is known as reformulated gasoline. New types of fuel formulations, more environmental-friendly than the regular fuel formulation, now contain additives like methyl t-butyl ether, methanol or ethanol, which are oxygen suppliers and reduce the emission of carbon monoxide during fuel combustion in an automobile engine. Legislation in different countries has been enacted to require the use of reformulated oxygen-enriched gasoline in locations of severe air pollution problems. Similarly, environmentalists are studying the effects of having vast areas of prime agricultural land to continuous production of corn, whose production depends on heavy application of synthetic fertilisers. Appropriate legislation in this field is urgently needed.

Source: *Environmental Chemistry. A Global Perspective* by G. VanLoon & S. Duffy. (2000). OUP, 83-88. By permission of Oxford University Press, Inc.

In this example, first the problem to be analyzed is stated: urban air pollution as one of the most worrying environmental problems nowadays. Then two main factors contributing to the problem are identified and finally some possible solutions are pointed out. Again, notice the expected overlap between two different patterns, problem-and-solution and cause-and-effect, since in a problem-and-solution pattern the causes having an effect on the problem are often examined.

Listing

Listing is a very common way of presenting information in technical and scientific writing. Lists may be either formatted or unformatted. Formatted lists are easier to identify since they are usually marked by means of indentation, letters or numbers. On the other hand, unformatted lists are not visually marked so it is especially important to use appropriate parallel structures so that readers can easily identify them. This pattern also includes listing by order of importance, which implies arranging the specifics from the least important to the most important or from the most important to the least important.

> ### Sample 1
> Transitions usually produce gradual changes in water prism cross sections and are used at structure inlets and outlets and at changes in canal sections to: (1) provide smoother water flow, (2) reduce energy loss, (3) minimize canal erosion, (4) reduce ponded water surface elevations at cross-drainage structures, (5) provide additional stability to adjacent structures because of the added resistance to percolation, and (6) retain earthfill at the ends of structures.
>
> Source: Aisenbrey. A.J, Jr, R.B. Hayes, H.J. Warren, D.L.Winsett & R.B.Young. (1977). *Water Resources Technical Publication.* United States Department of Interior. Bureau of reclamation, 335.

> ### Sample 2
> The engineer's chief responsibility is to ensure that the structure remains serviceable or fit for use throughout its design life. A structure that has become unfit for use may be said to have reached a *limit state*. There are many such states, the most important being collapse (structural failure); loss of stability, whether by overall tilting or by buckling of individual members; excessive deflexion; and excessive local damage (spalling of concrete or cracking, leading perhaps to corrosion reinforcement).

These two samples include lists in order to organize information. Nevertheless, whereas the list in the first paragraph is marked by numbering, the one in the second paragraph is unformatted. Also notice the use of parallel structures in both samples. In sample 1 the use of infinitive verbs contributes to binding the list together while in sample 2 the use of noun phrases facilitates the location of the four items included in the list. Moreover, also in sample 2, the list is arranged in descending order of importance, which has been clearly signalled by the author by introducing *the most important* with the first item. Finally note how both lists have been previously presented to make readers aware that a list is coming and to give them a tip about its nature. In sample 1 the sentence *"Transitions usually produce gradual changes in water prism cross sections and are used at structure inlets and outlets and at changes in canal sections to:"* tells us that the purposes of transitions are going to be described. With the sentence *"There are many such states"* in sample 2 we can easily guess that a list of such limit states will follow.

In this section we've seen some of the patterns of paragraph development that can be frequently found in technical and scientific writing. Nevertheless, there are also some others that could be included in this list such as for example logical division, extended definition and classification. Examples of these other patterns can also be found in different sections of this book. Besides, it should be noted that we can often find two or more patterns in combination. We have already seen in this section a *cause-and-effect* pattern that can also be interpreted as a *problem-and-solution* one or a *listing* in *order of importance,* just to mention a couple of examples. This overlapping is very common but no matter how patterns may be embedded, combined or interwoven, there should always be a pattern ruling most of the paragraph and providing a main structural guide.

INDIVIDUAL TASKS

3-10 Read the following paragraphs and decide on their pattern of organization. Also underline as many of their identifying features as you can. Remember that some of the paragraphs may develop more than one pattern.

Sample 1
As it is widely known, smog is an urban air pollution problem. There is a broad variety of types of smog depending on the local situation of every city or area, yet two general classes have been identified. Classical smog consists of carbon-based soot and other solid particulates and sulfur dioxide. Its most important source is combustion of coal and its most worrying property is that it has reducing and acidic features. The second kind of smog is photochemical smog, which basically consists of nitrogen oxides, ozone and other organic oxidants, and aldehydes. Photochemical smog oxidation reactions are initiated by the hydroxyl radical, which is mostly produced due to the presence of nitric oxide from combustion emissions. Engines of various kinds are the major factors responsible for the release of nitric oxide and carbon compounds, as they are essential to generating photochemical smog. The major sources of hydrocarbon emissions from gasoline-powered vehicles come from the crankcase (20 %) and from the fuel tank and the carburetor (15 %), while almost all emissions of carbon monoxide and oxides of nitrogen come from the exhaust pipe (65%).

Sample 2
At present, knowledge of plutonium and highly enriched uranium (HEU) inventories is incomplete and is largely kept under wraps by governments, industrial companies and international organizations. In companies possessing nuclear weapons or trying to acquire them, information about HEU and plutonium produced for military purposes is

generally classified. In the civilian context, information gathered by international agencies for safeguards purposes is held on a confidential basis and is not open to detailed public scrutiny, or even to the scrutiny of national authorities. The International Atomic Energy Agency (IAEA) and Euratom only publish broad aggregates so as to protect the identity of the countries and industrial operators providing the information. In all areas, whether military or civil, that information that does exist in the public domain is often inconsistent, scattered, and incomprehensible to the layman. Thus, there is need for greater transparency with regard to inventories of nuclear materials and this report, bringing together the information in an international register, should be updated in regular intervals to extend the Treaty on the Non-Proliferation of Nuclear Weapons (NPT).

Source: *World inventory of plutonium and highly enriched uranium by* D. Albright, F. Berkhout & W. Walker. (*1993*). OUP, 4-5. By permission of Oxford University Press, Inc.

Sample 3

A steel pot used to hold molten magnesium alloys leaked, releasing 80lbs of molten metal onto the foundry floor. Based on the evidence of the analyses performed, several hypotheses were formed which basically predicted that the oxide mass detected was introduced in the steelmaking process. The presence of the large oxide defect in the steel could cause the pot to fail in the following way: wear or erosion of the inner surface of the pot permitting the molten magnesium alloy to contact the iron oxide brought about a thermite-like reaction, which by its strongly exothermic nature would locally melt the steel, thus giving rise to the leak. The ensuing fire in the furnace hearth would overheat the outside surface of the pot bottom in a very unstable way, resulting in its rapid erosion and unusual microstructure.

Source: Tung, P., S. Agrawal, A. Kumar & M. Katcher. (eds.) (1980) *Fracture and Failure: analyses, mechanisms and applications*. Proceedings of the American Society for Metals Fracture and Failure Sessions. American Society for Metals: Ohio, 131-146.

Sample 4

It is the engineer's task to disentangle the complex situation of distinguishing the four main forms of wear from a few marginal processes that may often also be classified as forms of wear. The most common and least preventable type of wear is adhesive wear, which occurs when two smooth bodies are slid over each other and fragments are pulled off one surface and adhere to the other. The second most universal type in terms of the financial loss it produces is abrasive wear, which occurs when a rough hard surface, or a soft surface containing hard particles, slides on a softer surface and ploughs a series of grooves in it. Surface fatigue wear comes next, which is the wear that is observed during repeated sliding or rolling over a track. Finally, the most benign type of wear is probably corrosive wear, which appears when sliding takes place in a corrosive environment; in the absence of sliding the products of the corrosion will form a film on the surfaces.

Source: *Friction and wear of material by* E. Rabinowicz. Copyright © (1995 John Wiley), 128-133. Reprinted with permission of John Wiley & Sons.

Sample 5

The four-stroke cycle of an internal combustion engine is the cycle most commonly used for automotive and industrial purposes today. It was invented by a German engineer Nikolaus Otto in 1876 and, hence, is also called the Otto cycle. It is characterised by four strokes. The cycle begins at top dead centre, when the piston is at its topmost point. On the first downward stroke (intake) of the piston, a mixture of fuel and air is drawn into the cylinder through the intake (inlet) port. The intake (inlet) valve (or valves) then close(s), and the following upward stroke (compression) compresses the fuel-air mixture. The fuel-air mixture is then ignited, usually by a spark plug for a gasoline or Otto cycle engine, or by the heat and pressure of compression for a Diesel cycle of compression ignition engine, at approximately the top of the compression stroke. The resulting expansion of burning gases then forces the piston downward for the third stroke (power), and the fourth and final upward stroke (exhaust) evacuates the spent exhaust gases from the cylinder past the then-open exhaust valve or valves, through the exhaust port.

Source: Wikipedia, the free encyclopedia. [WWW page]. URL http://en.wikipedia.org/wiki/Four-stroke_cycle.

Sample 6

The four most important data on the chemical properties of solid wastes that need to be gathered are: 1) the proximate analysis, resulting from information about moisture, volatile matter, ash, and fixed carbon; 2) the fusing point of ash; 3) the ultimate analysis, resulting from the percent of carbon, hydrogen, oxygen, nitrogen, sulphur and ash; and 4) the heating value. Information about the chemical composition of solid wastes is important if we want to evaluate alternative processing and energy-recovery options.

Source: Peavy, H., D. Rowe & G. Tchobanoglous. (1985). *Environmental engineering*. New York: McGraw-Hill Co., 576-582. With permission of the McGraw Hill Companies.

3-11 Read the following paragraph and add appropriate chronological linking expressions in order to get a well-written paragraph using chronological or sequential order as its basic pattern of organization. You may also need to add punctuation signals.

The construction of a gravel beach is a solution within a wide range of alternatives, from an unprotected sandy beach to a concrete seawall, to reduce beach erosion. In the design process of a gravel beach (1)_____ the characteristics and the behaviour of the beach area in the past are studied. (2)_____ the ideal material for the new beach is defined, taking into account the rate of loss of material in the short and long terms. (3)_____ the availability of gravel including the production costs are studied. (4)_____ the new beach profile is designed, taking into account hydraulic aspects (overtopping, toe erosion, etc.), execution methods and costs are defined. In order to improve the stability of the beach, additional measures are considered. (5)_____ an optimisation of the beach design is done with respect to the type of gravel to be used, life expectancy, application of additional protection of the beach, risks of damage to the beach and consequent damage and costs.

Source: CIRIA/CUR. (1991). *Manual in the use of rock in coastal and shoreline engineering.* London: CIRIA/CUR, 454-455.

3-12 Identify the pattern of development of the following paragraph. Then add suitable signals and expressions where necessary or change some in order to improve the paragraph coherence and to emphasize the point the author tries to make.

Filtration processes are used primarily to remove suspended particulate material from water and are one of the unit operations used in the production of potable water. There are two main types of filtration: cake filtration and depth filtration. These two processes are very different. Cake filtration is the physical removal by straining at the surface. With this type of filtration, filtrate quality improves as the filter run progresses, and deterioration of the filtered water quality is not observed at the end of the filter cycle. In addition, chemical pretreatments are not generally provided. To obtain reasonable filter cycles the source water must be of quite good quality. Depth filtration involves complex mechanisms to achieve particulate removal. Transport mechanisms are needed to carry the small particles into contact with the surface of the individual filter grains, and then attachment mechanisms hold the particles to the surfaces. Chemical pre-treatment is essential to depth filtration. Rapid granular-bed filters are of the former type and precoat and slow sand filters are of the latter type.

Source: Pontius, F. W. (ed.). (1990). *Water quality and treatment: a handbook of community water supplies.* (4th ed.). *American Water Works Association.* New York: McGraw Hill, 455-457. With permission of the McGraw Hill Companies.

3-13 Write a chronological or sequential paragraph. Whether it is a recount of past events or a description of a process, remember to follow the steps below:

1. Write an outline including the different events organized by time.
2. Think of a topic sentence that names the process or that summarizes/lists the different events.
3. Develop your paragraph adding suitable chronological linking expressions. Also remember to include enough details so as to make your description clear.

3-14 Working as an engineer, imagine you have to write a paragraph that discusses one of the topics below in terms of cause-and-effect. First, note down the different points you will include. Then, think of a topic sentence to introduce the discussion and, finally, develop your paragraph adding suitable cause-and effect linking expressions. Short instructions are given to contextualize every situation.

Topic suggestions:
- Pollution (laypeople audience in a popular science publication; informative purpose)
- Friction in machines (executive audience in an introduction to a report; inform + report)

- The spread of the Internet (mixed audience in a university journal; inform + instruct + persuade)
- The advance of satellite communications (expert/student audience in a memo; inform + persuade)
- Construction materials (technicians audience in an in-company memo; instruct + inform)

(You may also choose any other topic of your interest related to your field of studies.)

COLLABORATIVE TASKS

3-15 Decide which pattern of paragraph development the following topic sentences suggest. Then, add two more topic sentences and let your partner guess the type of development implied.

1) Most large pumping stations abstract water from surface sources such as rivers, canals, lakes etc. whereas groundwater abstraction is usually provided by smaller pumping units.

2) The advent of relatively inexpensive cast iron and of wrought iron and the rapid spread of railroads in the mid-1800s gave rise to a golden age of bridge building.

3) The hydrologic cycle is a continuous process by which water is transported from the oceans and after following a series of stages goes back to the sea.

4) The use of roller compacted concrete (RCC) has changed the design and construction of new concrete dams and provided economical and multi-functional material for the rehabilitation of existing dams.

5) The main components of a rod extensometer are anchors, rods inside the protective pipe, and a reference head.

6) This report will compare laptop computers on the basis of the following: (1) features, (2) performance in hardware tests, and (3) price.

7) The main hazards of the process industries arise from the escape of process materials which may be inherently dangerous (e.g. flammable or toxic) and/or present at high pressure and high or low temperatures.

8) During the nineteenth century experimental hydrology flourished.

3-16 Sound waves travel through air in much the same way as water waves travel through water. Since water waves are easy to see and understand, they are often used as an analogy to illustrate how sound waves behave. In pairs, using the cues below, write an analogy on sound waves and water waves to a laypeople audience. Note that you may also have to use cause-and-effect linking expressions, at least for the first part of the paragraph, which is a description of sound waves. The figures below illustrate some wave properties in order to facilitate your understanding of the definitions and to help you develop your paragraph.

sound waves
- variations of pressure in a medium such as air
- created by the vibration of an object → this causes the air surrounding it to vibrate
- vibrating air causes the human eardrum to vibrate → the brain interprets this as sound

sound wave motion and characteristics analogous to water wave motion
- constant velocity of propagation
- waves have crests and troughs
- *wavelength*: the distance between any point on a wave and the equivalent point on the next phase
- *amplitude*: the strength or power of a wave signal. The "height" of a wave when viewed as a graph
- *frequency*: the number of times the wavelength occurs in one second. Measured in kilohertz (Khz), or cycles per second. The faster the sound source vibrates, the higher the frequency.

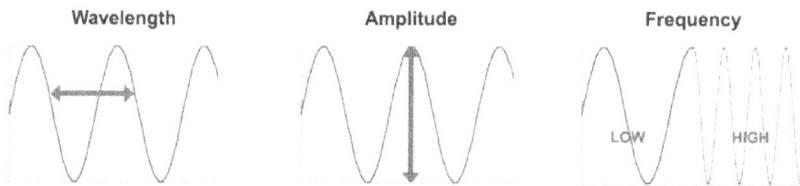

3-17 As you will learn in the next section of the book, these patterns of development are also applied when developing a whole text. Therefore, from the list of topics below decide what pattern of organization you might use. Compare your answers with those of your partner in order to see if she/he has different possible options you hadn't thought of.

1. Explain the "greenhouse effect" and describe alternatives to minimize its effects.
2. Discuss the advantages and disadvantages of different types of programming languages.
3. Briefly discuss the events leading up to the creation of the Internet.
4. Discuss the main applications of composites in industry. (You may also choose any other type of material you know well)
5. Describe how the advent of new world communications affects the life of the average person today.
6. Describe the hydrologic cycle.
7. Describe the main components of a computer.
8. Describe the main uses of a satellite.
9. Outline the three most important benefits of fibre optic communications.
10. Describe how the thrust of a jet engine resembles the thrust produced by a toy balloon when it is released after having been blown up.

3.5 Providing intra- paragraph coherence

Having covered the main techniques for proper paragraphing, we will now deal with coherence, which lies at the heart of good writing. Coherence plays a vital role in chaining ideas and improving the readability of texts, which implies writing sentences and arranging them into paragraphs to make up a sound and clear text. Essentially, this comes down to:

a) writing or combining sentences in an orderly and readable way *within* a paragraph, and

b) arranging and linking up ideas logically *between* paragraphs.

From the above explanation, it can be deduced that coherence may be viewed from two perspectives, *intra- paragraph coherence* and *inter- paragraph coherence*. If you stop for a moment and think of the difference between Internet and Intranet, you will probably rightly guess that *intra- paragraph coherence* refers to the ways sentences can be coherent *within* an isolated paragraph, while *inter- paragraph coherence* envisions the paragraph from a global perspective (different paragraphs that make up the text or document) and refers to how coherence can be achieved *between* paragraphs. Since this section is devoted to the paragraph, we will be looking at aspects that have a bearing on coherence *within* a paragraph, leaving those that provide coherence *between* paragraphs for section 3.8. By now you may be wondering: What exactly is coherence? The verb cohere means 'hold together', so intuitively you might guess that coherence constitutes those relationships which link the meanings of sentences in a text. In other words, coherence is the result of integrating sentences in the text. Sentences can make up a coherent text by at least three means:

1. shared knowledge between reader-writer.
2. unity of information or content.
3. cohesive devices, connectives, and other linguistic resources

Coherence, then, is an important factor to achieve readability; that is, to write documents which are easy to process and understand. When a text is readable, readers needn't stop to puzzle out the meaning or relationship between sentences. Besides, as form reflects content, it pays to transmit content in a coherent way. Having highlighted the positive effects of coherence, let us now delve into these three ways through which coherence can be achieved.

Shared knowledge

It is worth noting that the links integrating the meaning of sentences may be linguistic (grammatical or syntactic) or not. Note for example:

> The Sasser worm can infect our team. We should install an updated anti-virus program.

Even though there is no linguistic link, this text is coherent because the links are based on the communicators' shared knowledge about computers, viruses and so on. The writer/reader's knowledge on the topic and other contextual factors instruct readers to interpret this as a unified and integrated text. Here we find another reason that justifies the importance of identifying the kind of audience you are writing to; most importantly, it is crucial to know your readers' level of expertise and the knowledge you can assume they have on the topic. Now read the following:

> The Sasser worm, namely one of the latest computer viruses, can infect our team. Therefore, we should install an updated anti-virus program.

This text is also coherent. The writer has decided to be more explicit and defines *the Sasser worm*—assuming the reader needs being reminded or told. Besides, he has related the meaning of the two sentences by means of two connectors (*namely* and *therefore*).

Unity

The first and most basic means of writing a coherent paragraph is by *unifying content*; that is, by presenting and arranging similar and related content in a logical and orderly way. This can be achieved, for example, by using *patterns of paragraph development*. At other times the order in which information is presented is dictated to writers in such a way that they only have to conform to the standard or conventionalized format of a given genre. But, if this is not the case, then it's the writer's responsibility to organize information within paragraphs logically. A paragraph displays unity when it contains information that is both related and relevant to the topic and when it presents this information in a logical order. When unexpected off-topic information is added, the readers' attention is distracted from the paragraph main idea, which may cause misunderstanding.

Another way to ensure unity is by means of *parallel structures*. Parallel structures reflect similar kinds of information (at paragraph level and at sentence level) and can also act as content-unifying devices. Parallelism consists in using similar grammatical structures for similar ideas in such a way that items of equal importance are presented in the same grammatical form and used consistently. This can be seen in the examples below:

> (parallel) The materials are composed of concrete *reinforced with* steel bars and *pre-stressed with* steel wire, strand, or alloy bars.

> (nonparallel) The materials are composed of concrete reinforced with steel bars and steel wire, strand, or alloy bars are used to pre-stress it.

Coherent devices and other linguistic resources

Given that coherence mainly has an integrative function, linking expressions and other linguistic resources that connect sentences and help the reader interpret the message are regarded as items that contribute to coherence. Among other things, coherence-giving items can assist readers in: i) inferring meaning, ii) explicitly signalling the relationship between two or more sentences or ideas, and iii) creating a logical flow of sentences in a paragraph. More precisely, the following ways to achieve coherence will be tackled:

- coherence through the repetition of key words and through the use of pronouns and synonyms.
- coherence through connecting expressions.

Use of key words, pronouns, and synonyms

One way of achieving coherence is by *repeating key words* in order to emphasize the most important idea in a paragraph. Key words can be frontloaded and be used following the so called *Old-New principle* (also known as the *Given-New* information). This principle is based on the assumption that old information or information already known to the reader had better be placed first, before new information is conveyed. Since important or known information must be put first or put in important places, it is natural to place key words in a dominant position, that is, in subject position. When a key word is repeated in a dominant position, readers are able to go back to the same idea and find it further developed or elaborated in another way. By doing this, information is claimed to be processed better and faster, since old information is acting as a frame (sometimes as a reminder) to which new information is added.

Read the passage and then circle the pronouns that refer to the key word, *gasoline engines*:

> Gasoline engines are alike electric engines in their ability to pump water. They are also similar in that they need electric energy. In spite of being very cheap to install, gasoline engines have a high operating cost. Gasoline engines need batteries to start, and consequently they require constant electric supply. However, gasoline engines are very convenient for emergency situations, or in case of a blackout, as they basically derive their energy from fuel, not from electricity.

What is the antecedent of the two 'they' pronouns in the second sentence?

Note how less coherent the same passage becomes if the key word is only used once and is replaced with pronouns:

> Gasoline engines are alike electric engines in their ability to pump water. They are also similar in that they need electric energy. In spite of being very cheap to install,

> they have a high operating cost. They need batteries to start, and consequently they require constant electric supply. However, they are very convenient in emergency situations, or in case of a blackout, as they basically derive their energy from fuel, not from electricity.

However, key words should not be repeated too many times; it is advisable to make a balanced use of *pronouns* and *key words*. Also note that when a pronoun refers back to a key word it is said to be anaphoric and that the anaphoric use of pronouns does clearly provide coherence to sentences. In the paragraph below, see how the paragraph is improved by making a strategic use of pronouns and key words.

> **A programmable robot** is a sequential machine that correlatively executes the instructions indicated in the user's program and generates orders or control signals read at the entrance of the plant. When **the programmable robot** detects some change in the signal, **it** reacts according to the program stored in **its** memory until **it** obtains the orders required to quit. This sequence is executed continuously in order to obtain an updated control of the process. **A programmable robot** is useful to undertake tasks involving some control or transformation process; therefore the application of **programmable robots** ranges from industrial manufacturing and transformation processes to facility control.

Finally, important information can also be restated; if writers realize they have already repeated a given key word several times, then they may decide to use a *synonym* or a *substitution* of a key word instead, which also gives coherence to the text. However, writers should make a careful use of synonyms. An indiscriminate use of synonyms of key words may distract or even confuse readers to the extent that they may understand that a new and different topic is being dealt with.
Can you identify the synonym of the key word "cement" in the following passage?

> Portland cement or cement, as it is commonly known, is widely used nowadays in buildings, roads, bridges and dams. Cement consists chiefly of calcium and aluminium silicates. In fact, it is a mixture of compounds such as limestone, clay or shale. These are ground, blended, fused to clinkers in a kiln, cooled, and ground to the required fineness. The material is shipped in bulk or in bags weighing 94lb.

To conclude, we can say there is no rule as to the 'good' proportion of key words repeated and the use of pronouns substituting these key words. However, here is a sound principle: whenever meaning is not clear, repeat a key word instead of substituting it by a pronoun or a synonym, and mix repetition, pronouns and synonyms effectively in your writing. As a reminder, try to apply the *Old-New principle* throughout when repeating key words or using pronouns or synonyms. Remember that this strategy guides readers and gives them the context or frame as to what information they can expect to find when they read on.

Use of connecting expressions

The label "connecting expressions" encompasses a wide range of items like connectors, transitional signals, correlative structures, or links that can be used between sentences of different types. These devices play a major role in making a text coherent as they tend to ensure no sudden jumps between different ideas. They also help ideas flow smoothly and logically, anticipate what is to come and signal how one idea or sentence is related to another. Sometimes they refer to the text itself and its organization, and other times they reflect some kind of dialogue that is established between writers and audience.

Here is a table that summarizes the most common connecting expressions for general use classified according to their meaning and their syntactic features.

Connecting expressions / Meaning	Sentence connectors	Coordinating conjunctions	Subordinating conjunctions	Correlatives	Phrase linkers

To add information	in addition
	furthermore
	further
	also
	besides
	moreover
	what's more
	for one thing
		and		both ...and
				not only...
				    but also
				neither...nor
					together
					    with
					along with
					in addition
					    to
					as well as

To list or to introduce a conclusion	to begin with first and foremost first(ly) second(ly) third(ly) last(ly) next then in conclusion to conclude in summary to summarize in brief / briefly to sum up finally eventually last but not least				

To introduce a choice or an alternative	otherwise instead alternatively	or		either...or whether ...or	instead of
To restate	in other words rather namely viz. i.e. that is to say that is	or rather but rather			in other words or rather namely viz. i.e. that is to say that is
To exemplify	for example for instance e.g.				for example for instance e.g. like such as including to illustrate to exemplify
To introduce contrast or concession	however yet still nevertheless nonetheless on the other hand on the contrary by/ in contrast, all the same	but yet	although even though though while whereas		in spite of despite unlike in contrast with as opposed to
To introduce a cause or result	therefore thus[6] as a result as a consequence consequently that is why for this reason hence[6]	so for	since as because for the reason why given that		because of due to owing to as a result of on account of the reason for as a conseq. of

---

[6] **Thus** and **hence** can also be used as phrase linkers. In such cases, **thus** is preceded by a comma and followed by a gerund: *He painted the metal with a protective paint, thus (thereby) avoiding corrosion.* **Hence** is typically followed by a noun: *This book on the strength of materials is a comprehensive, didactic and updated handbook on the field, hence its success.*

To introduce purpose			in order to so as to so that		

To introduce a condition			if-clauses
			unless
			in case
			provided (that)
			so long as
			on condition
			   that
					in case of
					in the event
					   of

To introduce time	after that
	afterwards
	subsequently
	in the
	   meantime
	meanwhile
	simultaneously
	at the same
	   time
	then
	next
	later
	at this point/
	   stage
	previously
			before
			after
			when
			whenever
			until /till
			since
			once
			as
			while
			as soon as
					before
					after
					until
					when /on
					/in
					previous to
					prior to
					following

To emphasize attitude and certainty	in fact
	really
	actually
	as a matter of
	   fact
	indeed
	of course
	surely
	certainly
	obviously
	definitely
	hopefully
	to some extent
	strictly
	   speaking
	essentially
	theoretically
	sincerely				

	personally to be precise				

Table 3.1 Connecting expressions.

These connecting expressions cannot be used indistinctively. Some of these devices are used in simple sentences, others are found in complex or compound sentences and, most importantly, they all require different **punctuation**. Given that having a good command of these connecting devices implies punctuating them correctly, you should carefully study the figure below and then read the examples that follow.

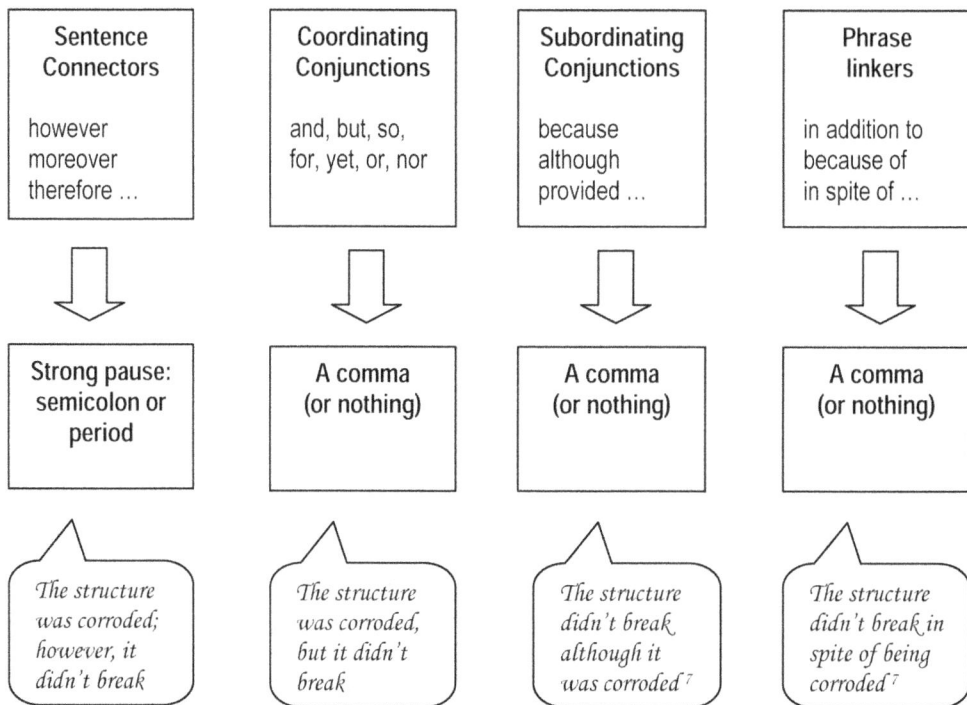

Fig. 3.2 Punctuation of connecting expressions

---

[7] When the subordinated clause, introduced by the subordinating conjunction or phrase linker, precedes the main clause then the use of the comma is necessary.

  e.g. Although it was corroded, the structure didn't break.

  e.g. In spite of being corroded, the structure didn't break.

Sentence Connectors join two independent sentences, they usually appear at the beginning of the sentence and they are always followed by a comma. Though not so common, they can also appear in the middle (between commas) or at the end of sentences.

> e.g. He was a supporter of the project; however, he wrote a report that constituted an attack on the engineering firm in charge.
> e.g. He was a supporter of the project, but he wrote a report that constituted, however, an attack on the engineering firm in charge.
> e.g. He was a supporter of the project. He wrote a report that constituted an attack on the engineering firm in charge, however.

Coordinating Conjunctions are used to join two independent sentences and are normally preceded by a comma.

> e.g. He wasn't a supporter of the project, so he wrote a report that constituted an attack on the engineering firm in charge.

Subordinating Conjunctions are used to introduce a dependent clause. This dependent clause they introduce can be placed before or after the independent sentence to which it is subordinated (note the use of the comma when the dependent clause comes first in the sentence):

> e.g. Although he was a supporter of the project, he wrote a report that constituted an attack on the engineering firm.
> e.g. He wrote a report that constituted an attack on the engineering firm although he was a supporter of the project.

Correlatives are placed before the elements they join in the sentence, and they are a good means to achieve parallelism, and therefore coherence.

> e.g. Advances in technology are *both* necessary *and* advantageous for many people.
> e.g. Advances in technology are claimed *not only* to raise our standard of living *but also* to create an even deeper trench between rich and poor countries.
> e.g. If technology doesn't go hand in hand with globalization, the different countries will have neither the same opportunities nor the same development.
> e.g. Advances in technology may have *either* positive *or* negative effects depending on their rational use.
> e.g. It remains to be seen *whether* advances in technology will necessarily help the human race progress *or* not.

Phrase linkers are used to introduce a noun phrase or a nonfinite clause (present and past participle or to-infinitive)

> e.g. *Because of* discrepancies with the board of directors, he wrote a report that constituted an attack on the engineering firm in charge of the project.

e.g. *Despite* being a supporter of the project, he wrote a report that constituted an attack on the engineering firm in charge.

e.g. He had a strong argument about the feasibility of the project with the board of directors. *In addition to* this, he wrote a report that constituted an attack on the engineering firm in charge of the project.

So far we have been dealing with the use of connecting expressions. However, it is also very important to make an adequate and varied use of them. In this way, writers may be able to find a compromise between always writing short, simple, telegraphic sentences (which would make your text sound childish and reflect a poor command of English), and always writing long, complex and wordy sentences (which would not comply with the conciseness and clarity features of technical English). Try to find a happy medium in the quantity and degree of formality of connecting expressions you use.

There are other reasons that highlight the importance of using connecting expressions. First, since form reflects content, poorly written documents may well reflect a poor content, too. Second, the English rhetoric in general tends to be concise and direct; English is a synthetic language as opposed to more rhetorical languages where more subordinated sentences abound. Finally, acquiring some fluency and expertise in using connecting expressions properly and when necessary will be a must for a non-native student of technical written communication. Compare these different versions of the same short narrative and decide which one you prefer:

Sample A

The fittings in the distributing pipe system were oxidised. They lacked maintenance. A T-junction joint was leaking. It had to be replaced. There was a chisel in the toolbox kit. He took it. It was the first time he tried to replace a T-junction joint. He took the wrong tool. He damaged the joint. Another tool was chosen, a pair of flat-nose pliers. The operation was repeated. He succeeded. The joint was replaced.

Sample B

Since the fittings in the distribution pipe system were oxidised **due to** a lack of maintenance, there was a T-junction joint **which** was leaking **and** had to be replaced. There was a chisel in the toolbox kit **so** he took one. **However, as** it was the first time he tried to repair a T-junction joint he took the wrong tool and damaged the joint. Another tool was chosen: a pair of flat-nose pliers. **When** the operation was repeated, he succeeded in replacing the joint.

Sample C

Given that the fittings in the distributing pipe system were oxidised **on account of** a lack of maintenance, a T-junction joint was leaking **which** had to be replaced. **Albeit** it was the first time he tried to replace a T-junction joint, he took a chisel from the toolbox, **which** proved to be the wrong tool to repair such a joint, **thereby** damaging it. **Subsequently**, another tool, **namely**, a pair of flat-nose pliers, was chosen and, repeating the operation, the joint was eventually replaced, and **so** the operation was a success.

The first passage is, no doubt, the least coherent of them all. It is choppy and it seems as though the writer does not have a good command of English and, therefore, does not want to run any risks. Although there are no grammatical mistakes, this passage is telegraphic and seems to have been written by a child rather than an adult. On the other hand, the writer of the third passage has abused connecting expressions. There is an overuse of subordinated and complex sentences as well as connecting expressions, which hinders comprehension. In brief, the passage sounds 'un-English'. On the contrary, the second passage is closer to the aforementioned happy medium, in which the balance between extremely short sentences and too many long subordinated sentences has been achieved.

CRITICAL THINKING

3-18 Look at the first passage of the examples above. As it is not a highly technical subject, you certainly understood the message, which is coherent even without any connecting expression. For example, what is the relationship established between the sentences "It was the first time he tried to replace a T-junction joint. He took the wrong tool"? Did you understand this relationship? How is coherence achieved between the two sentences?

INDIVIDUAL TASKS

3-19 Read the text carefully and choose the connecting expression that best fits the text. Only one option is correct.

Video Recording

For the last two decades digital technology has slowly invaded television -most of videotape electronics is digital. (1) ...., the recording process itself remains analog for three basic reasons.
The first is standardization. The recorded signal format on tape has to be standardized (2) .... a tape recorded on one machine may play back on a different machine. Program exchanges, including the distribution of commercials, rely heavily on standardized videotape formats. (3) ...., the VHS format has become the standard for videotape rentals (4) .... there are better-quality formats around.

A second factor is recording density and cost. Digitized video must cram tape (or any recording media) much more tightly with data than is the case with analog data. Analog video is typically imposed on a frequency-modulated carrier requiring 15-18 MHz bandwidth for professional recorders, (5) ………. uncompressed digitized video would need 30-100 MHz per recording head—and there are often many heads stacked in parallel. (6) ………., (7)………. recording digital data on tape is little harder than recording analog signals, digital videorecorders and magnetic tape both cost more, reflecting the higher price and greater number of high-speed large-scale ICs required and the expensiveness of the new, higher-density magnetic tapes.

A final barrier is the largely analog infrastructure of present-day television studios and production facilities. To accommodate digital recording fully, most installations should be replaced. (8) ………, digital pieces of equipment like professional videocassette recorders have become digital islands with analog inputs and outputs in an analog world.

Reprinted with permission from Hamalainen, J. Video Recording. *IEEE Spectrum*, April 1995, 76. © 1995 IEEE.

1. a) Despite   b) Therefore   c) However   d) Besides
2. a) so that   b) owing to   c) as though   d) while
3. a) But   b) Otherwise   c) Hence   d) Provided
4. a) and   b) even though   c) therefore   d) because
5. a) since   b) despite   c) whereas   d) so
6 a) In addition   b) On the contrary c) Conversely   d) Similarly
7. a) unlike   b) as a result of   c) in spite of   d) although
8. a) Finally   b) And   c) However   d) As a result

3-20 Read the following passages and fill in the blanks with suitable connecting expressions. Note that you can't make any changes in the passages below and that you can't use SO, BUT, AND, IF, RELATIVE PRONOUNS, or repeat connectors within the same passage.

Corrosion-resistant Materials and Coatings

Carbon and alloy steels do an excellent job of providing low to high strength fasteners at low moderate prices. (1)………….., they dominate the choices in fastener materials. (2)………….., they will rust in normal atmospheres and will corrode in more corrosive environments, which is a great disadvantage. (3)………….., corrosion-resistant alloys and coatings have been developed. Many of these are applicable to fasteners. Some, (4)……..…… their high nickel and/or chromium content, are also used at elevated and cryogenic temperatures.

Initially, corrosion-resistant fasteners were made of pure non-ferrous metals (5)………..… copper. Later, the addition of zinc produced the brasses, and the further addition of tin created the bronzes. Nickel and nickel alloys were to follow. Titanium came along in the 1950s after a process to produce it was developed. (6)……….….. titanium is often used where high strength and low density ((7)………..…, a high strength/density ratio) are required, it happens to be one of the most corrosion-resistant materials in the world. (8)……..…….. their good corrosion resistance, the non-ferrous and stainless steel fasteners could not match the strengths of the existing alloy steel bolts, and so the fastener manufacturer was challenged to provide high strength corrosion-resistant bolts and nuts. (9)............, it should be noted in passing that bolts and nuts are usually made of the same material (10)............. to provide corrosion compatibility.

Excavation: Site Preparation

Site clearance involves both the demolition of trees, shrubs, hedges, etc, and the demolition of existing structures on the site. Demolition is a specialist operation and, (1)............, should not be attempted by the builder on anything but the smallest outbuildings. By employing a reputable demolition contractor, there are many advantages to be gained: (2)..............., he can salvage the material and will have ready outlets for it, (3).............. reducing the cost of the operation; (4)............ he is going to achieve more safety, and (5).............., he will also have adequate insurance cover (6).............. any mishaps. Of all the advantages, safety is the all-important factor, (7)............. it applies (8)............ to the workforce but to the general public, adjoining property, and services as well.

3-21 This informal passage is written in a telegraphic style. Rewrite it and make the connections between sentences clearer by adding suitable links or making other changes you might consider necessary to improve the passage. Notice that there may be several possible alternatives:

GPS—a Global Navigation System Everyone Can Use

Traditionally, navigation was esoteric science. Someone got fed up and said: "That's it! We've got to have a system that works". That someone was the US Department of Defense. It was a massive undertaking. They had the money (over $12 billion) it took to do the system right. They came up with something to simplify accurate navigation. It was called the Global Positioning System or GPS. It's based on a constellation of 24 satellites orbiting the earth at a very high altitude. The satellites are high enough. They can avoid problems encountered by land-based systems. The satellites use technology accurate enough. They give pinpoint positions anywhere in the world, 24 hours a day. It is possible to get measurement accuracies better than the width of an average street. GPS was first and foremost a defense system. It's been designed to be impervious to jamming and interference.

What's most exciting is its potential. With today's integrated circuit technology, GPS receivers are fast becoming small enough and cheap enough to be carried by anyone. Everyone will have the ability to know exactly where one is, all the time. Knowing where you are is one of man's basic needs. This new service will become as basic as the telephone. A 'new utility'. That's just the start. Its applications are almost limitless. GPS allows every square meter of the earth's surface to have a unique address. New ways of organizing our work and play will be possible. Imagine a future. In a future a phone book is no longer a paper book. The phone book is a computer database in the memory of your computer. The database stores the exact GPS location of everything. You're looking for a Chinese restaurant; your computer can search through the phone database, find the location nearest to your current location and direct you to it immediately. No more aimless hunting. No more wasted driving.

Source: Hurn, J. (1989). *A Guide to the next utility*. Sunnyvale, CA: Trimble Navigation, 7-11. Reprinted with permission of Trimble Navigation. Ltd.

3-22 Organize the sentences into paragraphs according to the topics given below. Provide the necessary links to relate the sentences semantically and make any necessary changes to join the sentences coherently. Use the topics provided to write a topic sentence for each paragraph. Note that the sentences are already given in the right order.

How Microwave Cooking Works

PARAGRAPH 1: Advantages of microwave ovens
PARAGRAPH 2: The working principle: the microwaves
PARAGRAPH 3: How microwave ovens cook

1. The microwave oven is one of the great inventions of the 20th century.
2. Millions of homes in America have a microwave.
3. Microwave ovens are popular.
4. Microwave ovens cook food incredibly quickly.
5. Microwave ovens are extremely efficient in their use of electricity.
6. A microwave oven heats only food, nothing else.
7. A microwave oven uses microwaves to heat food.
8. Microwaves are radio waves.
9. Microwave ovens commonly use a radio wave frequency of roughly 2,500 megahertz (2.5 gigahertz).
10. Radio waves in this frequency are absorbed by water, fats and sugars.
11. Radio waves are absorbed.
12. Radio waves are converted directly into atomic motion -heat
13. Microwaves in this frequency are not absorbed by most plastics, glass or ceramics.
14. Metal reflects microwaves.
15. The metal pans do not work well in microwave ovens
16. In a conventional oven the heat has to migrate by conduction from the outside of the food toward the middle.
17. In the microwave cooking the radio waves penetrate the food.
18. The radio waves excite water and fat molecules pretty much evenly throughout the food.

19. With microwave ovens, heat does not have to migrate toward the interior by conduction.
20. Heat is everywhere all at once.
21. Molecules are all excited together.
22. There are limits.
23. Radio waves penetrate unevenly in thick pieces of food.
24. Wave interference causes "hot spots".
25. The whole heating process is different.
26. Microwave ovens excite atoms.
27. Conventional ovens conduct heat.

Adapted from: Brain, M. (April 01, 2000). How microwave cooking works. [WWW page]: URL http://home.howstuffworks.com/microwave.htm). (February 03, 2007). Courtesy of Howstuffworks.

3-23 All these sentences contain at least one nonparallel structure. Identify the nonparallel structure and then provide a more coherent version:

1. Both a brief review of material behaviour and describing the types of bar reinforcements are necessary.
2. You should apply a layer of this oil and the operation should be repeated twice in order to protect the bar so that it doesn't rust. (two changes)
3. The hinges squeak because they should be oiled and the cogs are rubbing as they should be oiled too.
4. After carefully evaluating the incident, we decided that we either have not the authority nor the means to cope with the problem.
5. This year many students have financial problems because of the rise of university fees and because fewer scholarships have been given by the government.
6. If you are to attend an interview, you had better wear smart formal clothes, arrive on time, and don't slump into the chair.
7. As an engineer, you are expected to be a productive worker, a creative thinker, and communicate efficiently.
8. By taking action to thwart global warming, companies can not only reduce costs but technological innovation can also be sparked.
9. Finding new sources of energy and the development of economical means to convert natural sources of energy into usable energy are a major challenge in the 21st century.
10. The article sets out the reasons why business is taking global warming so seriously and it is explained how companies are preparing for a carbon-constrained world.
11. To combat climate change, scientists, governments and business must act fast, they must unleash the talent inside business, and technological innovation must be pioneered.
12. The lack of political will and public knowledge seem to be the main reasons for the inaction, but we have two choices—more concern about global warming or, if not, we should be prepared to face the consequences.

COLLABORATIVE TASK

3-24 Imagine you have been asked to write a paragraph about mouse technology to an audience of lay readers with the purpose of informing them. Below you will find mixed information concerning different aspects of mouse technology. Read it and select the content you need depending on the approach you want to give to the topic. Also decide which visual aids may help you improve your description and refer to this figure or table within the text. After writing your paragraph, exchange it with a peer and assess each other's version using the assessment sheet below.

1. When mouse technology is patented
1970. Patent of an X-Y position indicator to move a cursor. (Stanford Research Institute).
1974-1975. One of the first patents to detail the use of a ball and optical signals. (Xerox Corporation).
1986. No need for a specific work surface. (California Institute of Technology).
1988. Early patent of a wireless mouse. (Mitsuboshi Belting).
1989. Ergonomic design introduction to reduce hand muscle fatigue. (Private inventor).
1994. Finger tip control of cursor for laptops. (AST Research).
1999. Mouse with a "rotatable and depressable" scroll wheel (Microsoft).
2005-06. Mouse integrated with an electronic massage function, with a telephone function (CAO & Web.de respectively) and later with a vacuum cleaner, a camera, or other functions. (Private inventors).

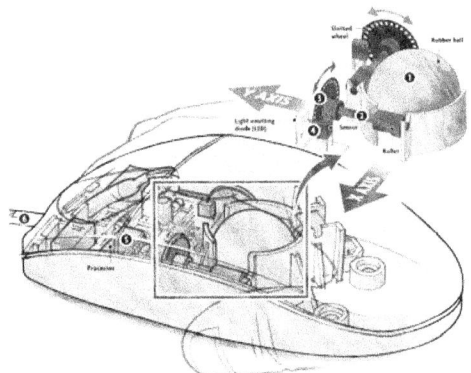

Fig.1

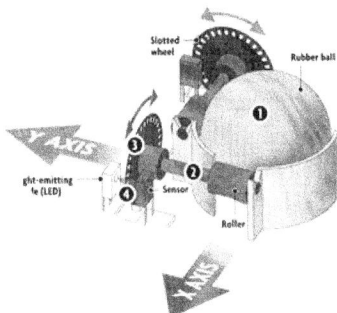

Fig. 2

2. Mouse description
Trackball mouse: a protruding rubber ball rolls / next to the ball, 2 rollers connected to 2 slotted wheels / each wheel, between a light-emitting diode & a sensor / light through no. of slots, distance indicated / speed of light bursts = speed of the mouse / one wheel indicate movement along X axis, the other along Y axis / the mouse processor interpret signal to plot position / signals sent, mouse position shown.
Optical mouse: no rubber ball, but a LED / LED flash over 1,000 times-second / sensor record patterns of light / processor calculate mouse movement.

3. Where mouse technology is patented

2000	No. of patents	2006	No. of patents
Japan	112	US	204
US	51	China	65
Germany	37	Taiwan	59
China	14	Germany	53
UK	6	Japan	41
S. Korea	3	S. Korea	14

Table 1

Assessment sheet

Questions:

1. Is there a good topic sentence?
2. Are all sentences related to the topic sentence? Can you spot any off-topic sentence(s)?
3. Is the paragraph pattern clear and consistent?
4. Is the paragraph coherent? Are sentences properly connected?
5. Is the content clearly and intelligibly transmitted? Are ideas well-expressed?

Question	Feedback	Grade
1		
2		
3		
4		
5		
	TOTAL	/10

3.6 Structuring the essay

We learnt that a paragraph is a linguistic unit used to develop a simple idea or topic. When this idea or topic is more complex, the paragraph becomes inadequate, insufficient; it becomes a too small linguistic unit to fit all the information in a clear and comprehensible way. In this case, we need a larger linguistic unit, one in which the main points and subpoints can be developed and clearly distinguished and organized in different paragraphs or sections. This larger unit can be any type of written text about a topic that has more than a paragraph and displays a structure. From now on, we will be using the term "essay" to refer to this larger unit. In this sense, an essay is similar to a paragraph in that

- it deals with a topic
- it develops a main idea through supporting facts (subtopics)
- it has a structure: an introduction, a body and a conclusion

However, an essay is longer and more complete than a paragraph because the topic of the essay contains subtopics which call for individual development in separate paragraphs. Thus we can say that *an essay is a group of paragraphs dealing with a topic which are headed by an introduction and closed by a conclusion.* It is important to point out here that the standard three-section structure of the essay, although usually found in most of the documents used in the technical and business-administrative fields, can appear with some variations, mainly with different title, different order, different size or different type of information. For instance, the information in a letter is organized in three main parts: the opening or introduction (where the writer introduces the purpose of the letter), the body (in which some supporting information is presented) and the closing (where the writer thanks or says farewell to the reader). However, a letter never ends with a conclusion as it would not comply with the conventional ending of this type of document: the purpose of the closing of a letter is not to reach any conclusion, so with letters there is an obvious change of information in the last section. Another interesting example in which a section has been moved is found in the administrative reports where the conclusion section appears right after the introduction and method instead of at the end of the document, as is the case with the vast majority of other types of documents.

The structure of the essay

As pointed out before, the information in an essay is organized in three main parts or sections, each of which has a specific purpose:

The introduction opens the essay. It is normally one paragraph long[8] and is made up of some *general information* and a *thesis statement*. The thesis statement is similar to the topic sentence of a paragraph as it introduces the main topic of the essay and names the major

---

[8] The introduction length depends on the text length. A long text may require more than one introductory paragraph.

subtopics that will be developed in the body (see figure 3.3 for a parallelism between the paragraph and the essay). However, the thesis statement is the last sentence of the paragraph.

The body follows the introduction and consists of one or more than one paragraph, each of which develops a subtopic. A parallelism can be drawn between the body paragraphs at the essay level and the supporting sentences at the paragraph level as each paragraph develops a subtopic of the essay in the same way as supporting sentences develop the topic of the paragraph.

The conclusion is the last part of the essay and usually has one paragraph. It is used to close the essay and has the same purpose as the concluding sentence in a paragraph (i.e. to restate the main idea or emphasize an important point), but with the main difference that the concluding sentence is more optional than the concluding paragraph.

Another important element to consider when writing an essay is the use of connecting expressions between paragraphs. Just as we used connecting expressions to join sentences within a paragraph in order to give unity and flow to the text so we need connecting expressions to tie up all the paragraphs in an essay to form a unified text. We will deal with this and other related aspects in the section on inter- paragraph coherence.

ESSAY

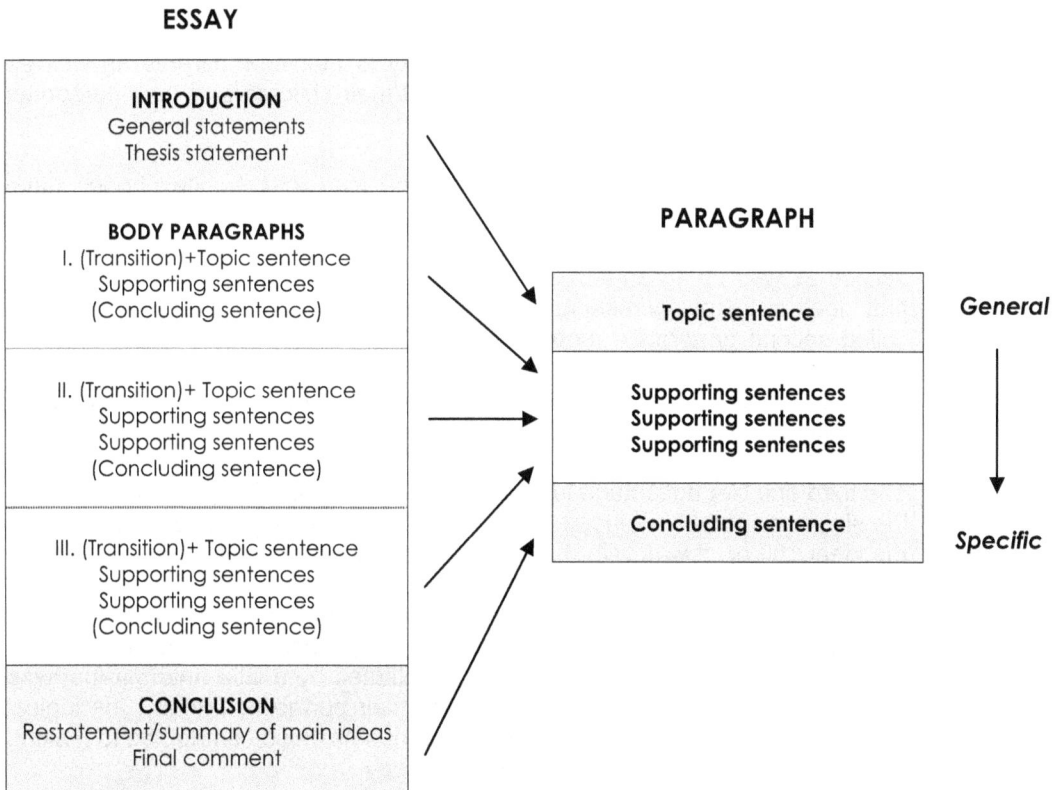

Fig. 3.3 Parallelism between the paragraph and the essay

Here is a sample of an essay that follows the introduction, body and conclusion organization.

Advantages of Mobile Phones

Introduction

It is an undeniable fact that mobile phones have shaped modern lifestyle. In the early 90s one could hardly imagine that those tiny devices would be everywhere in less than a decade. Continuous technological development has led to new methods of production, and consequently to lower consumer prices, which together with the marketing done by retailers have created the "mobile phone hype" as it exists today. However, the focal point that has made mobile phones irresistible to users of all ages is the appealing advantages they display.

Body

The first and foremost advantage is reachability. With a mobile phone you can communicate worldwide with anybody anywhere. This possibility has resulted in a high efficiency due to time and cost savings, as it is possible to fill in travelling time with working hours. In today's consumer society every second matters. Therefore, a short response time is crucial for both economic and business decisions. Besides, travelling safety has greatly improved since, in case of an emergency, it is possible to call for professional help immediately. One more advantage related to reachability is that now parents as well as bosses can communicate, or rather, control their children and their personnel respectively 24 hours a day.

The second advantage revolves around the idea that mobile phones have turned into something else than a pure communication device. The increased bandwidth has made it possible for mobile phones to become an information service as well as an entertainment centre, offering a long list of applications that does not stop increasing as new models of mobiles appear. If the so called second generation mobile phones (2G) incorporated a voice mailbox, the e-mail service and the SMS among others and represented a great step ahead in mobile communication, the 3G has gone far beyond by offering the possibility of taking pictures, videoconferencing and listening to music.

The third and last advantage is related to their technical features. Not only was the size considerably reduced to fit in one's hand but also the weight to reach the ideal 100 gr. It was with the discovery of semiconductor materials and the development of integrated circuits that technical experts were able to make this substantial reduction, which converted mobile phones into such an easily portable device. To this, the wireless technology used must be added. The cable connection of fixed phones was substituted by a little antenna that was integrated into the casing, thus facilitating their portability. Finally, the logical and easy-to-use distribution of keys and menus has also contributed to making this device a manageable gadget for everybody.

Conclusion   In conclusion, it is reasonable to say that the great impact that mobile phones are having on our society is mainly due to their advantages. Mobile phones have made it possible not only to call somebody from almost anywhere thanks to their wireless technology but also to obtain information from the Internet or have fun playing games. Yet, in spite of the advantages, we cannot forget that, as with most things, mobile phones have also some disadvantages that cannot be underestimated.

The introduction

The introduction is a very important part of the essay if we consider that this is the reader's reading start point: the reader may decide whether to go on reading or not depending on how interesting and useful he/she finds the introduction. A good way to organize information to obtain an effective introduction is to begin with *general information* about the topic and so provide some background to the reader and also arouse interest. This background information may have different purposes and therefore different forms in different documents: it can be a summary of the related literature, a description of analogous products in the market, or an explanation of the state of the art. The introduction, then, should proceed with more specific information that leads into the *thesis statement*. Therefore, the introduction will follow a general to specific pattern as depicted in the figure below:

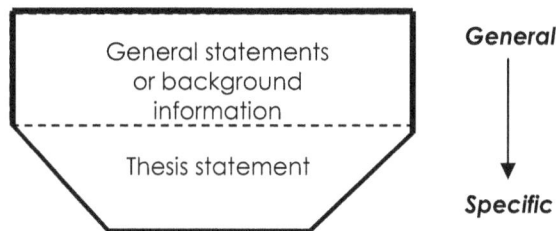

Fig. 3.4

The *thesis statement* is the statement that serves to introduce the main topic (or thesis) and the subtopics of the essay; in other words, it tells what the body paragraphs will be about. So by knowing the subtopics one can deduce the number of paragraphs in the essay. The thesis statement may be more or less specific; it can name the subtopics in a more general or specific way by outlining the subtopics and it may reflect the pattern of organization of the body paragraphs (e.g. comparison and contrast, order of importance, chronological order, etc.) but it can never be a too general or vague statement. Thus, we can define a thesis statement as a statement that:

- is the most important and specific in the introduction
- is placed at the end of the introduction
- introduces the topic of the essay

- may state the subtopics (which will be later found in the topic sentences of the body paragraphs)
- may indicate the pattern of organization of the body
- may be restated again in the conclusion
- is usually expressed in one sentence

The figure below graphically shows how the thesis statement unfolds within an essay.

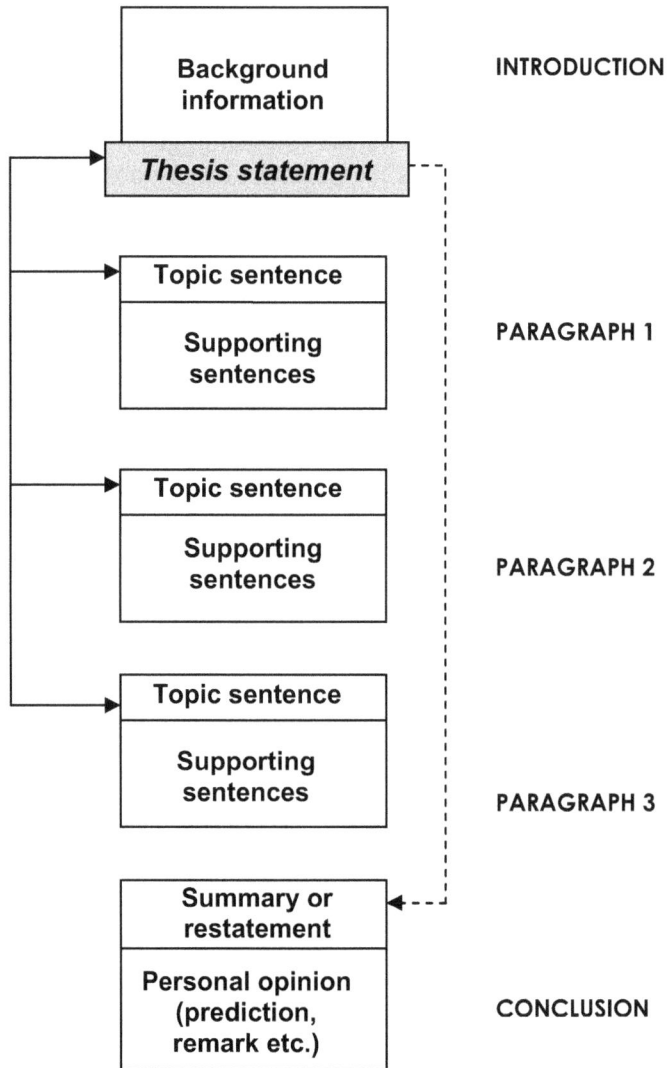

Fig. 3.5 The role of the thesis statement

Notice how all these features appear in the following introductory paragraph taken from the previous essay on the advantages of mobile phones.

> It is an undeniable fact that mobile phones have shaped modern lifestyle. In the early 90s one could hardly imagine that those tiny devices would be everywhere in less than a decade. Continuous technological development has led to new methods of production, and consequently to lower consumer prices, which together with the marketing done by retailers have created the "mobile phone hype" as it exists today (BACKGROUND INFORMATION). However, the focal point that has made mobile phones irresistible to users of all ages is the appealing advantages they display (THESIS STATEMENT).

This introduction has two clearly distinguishable parts. One part comes first and provides general information through several sentences about the importance of mobiles in modern life and the reasons for their wide and fast use. The subsequent part, the thesis statement, states the topic of the essay. It is worth pointing out here that in this particular introduction the thesis statement is expressed in general terms, since neither the subtopics nor the pattern of organization is indicated, although the essay is likely to follow a listing pattern (the appealing advantages will probably be enumerated and described in the body). A more specific thesis statement with the subtopics stated could be:

> However, advantages such as their enormous reachability, their wide range of applications and their incredible ease of use are what really make mobiles irresistible to users of all ages.

The previously described introduction is mainly found in academic texts or formally-written texts; texts in which the organization of information is a major concern and where an explicit indication of how information is organized (thesis statement) is required. Some examples of this type of texts are academic essays, descriptions, examination reports, progress reports, and thesis introductions. But not all documents will have the same type of introduction; some documents such as letters, memos, and proposals can have a shorter introduction where just one or two sentences introducing the purpose of the document will constitute the introduction. These purpose sentences, however, will not necessarily function as a thesis statement since on many occasions they do not indicate what will be discussed in the following paragraphs or sections –the structure of certain documents does not make a thesis statement necessary. On the other hand, longer documents such as reports have longer introductions containing more than one introductory paragraph, one of which will generally state not only the purpose of the document but the thesis statement as well. An example of this can be found in introductions of long reports where the writer devotes one paragraph to announce how many and what sections he/she will be dealing with. So in view of the diversity of introductions that different documents might display, the writer will be compelled to stick to the model agreed upon in a particular type of document. Consider the following progress report introduction:

> The purpose of this report is to inform you of the progress I have made on my final project since our last meeting on 19 July 20XX. This report will cover the work I have

> completed to date, the work I have yet to complete, and the problems I have encountered with the simulation work.

The previous example clearly shows how the purpose statement and the thesis statement are combined to make up the brief introduction of a progress report.

Different ways to begin an introduction

There are several possible ways to begin writing the introduction efficiently. Among the most common we find in the technical and scientific fields are:

1. *A paragraph consisting of some background information and a thesis statement.* The most common option is to initiate the paragraph with information that serves as a background explanation to the main topic before properly stating the main idea or thesis statement. This is usually found in research articles, academic essays and textbooks.

   > Every day, in our work and in our leisure time, we are in contact with and we use a variety of modern communication systems and communication media, the most common being the telephone, the radio and the television. Through these media we are able to communicate (nearly) instantaneously with people on different continents, transact our daily business, and receive information about various developments and events of note that occur all around the world. Electronic mail and facsimile transmission have made it possible to rapidly send and receive written messages across great distances. Can you imagine a world without telephones, radio and TV? Yet, when you think about it, most of these modern-day communication systems were invented and developed during the past century. (BACKGROUND INFORMATION) Below, we present a brief historical review of major developments within the last two hundred years that have played a major role in the development of modern communication systems. (THESIS STATEMENT)

   Source: Proakis, J & M. Salem. (1994). *Communication Systems Engineering*. Prentice-Hall Inc., 1. Reprinted with permission of the publisher.

2. *A purpose statement introducing the main objective or intention of the document.* Certain documents require an introduction opening with a straightforward purpose statement as it is the case with letters, abstracts and some type of articles.

   > We thank you for the letter of 10th August in which you draw our attention to an apparent error in the discount we have calculated on your quarterly statement dated 31 July.

3. *Some numerical and/or graphical information.* The description of a mechanism, a circuit, a graph and even a formula can also be a clear and illustrative way to introduce the reader to the text. This form can be mainly found in different types of articles.

   > The mobile telephone equipment consists primarily of a Base Band part, an IF part, and an RF part. Figure 1 shows a block diagram of the RF part, in which the duplexer

introduced in this article is employed. The duplexer is needed as the same antenna is used for both transmitting and receiving at the same time. It has two impedance-matched filters, one of which is tuned to the transmitting frequency (ft) and the other to the receiving frequency (fr). They are connected to the transmitting circuit and receiving circuit, respectively. (BACKGROUND INFORMATION) The desirable features of the duplexer are as follows: (THESIS STATEMENT)

Reprinted with permission from Komazaki, T. & K. Gunji. (1990). Attenuattion Pole Type Dielectric Filter for Duplexer. *40th IEEE Vehicular Technology Conference*, 59. © 1990 IEEE.

4. *An anecdote or some sort of surprising information.* This is a good way of attracting the reader's interest and finding out what the paragraph or text is about, and thus identify the main idea. Texts or documents aimed at a lay audience tend to use this form in order to catch the readers' attention. The example below illustrates how an anecdote may serve as an introduction to a technical topic. Also note that the whole story could be considered background information and that there is no thesis statement.

About 15 years ago, on a clear dark night on the Minnesota prairie, a young scientist testing his auroral imaging camera discovered giant flashes of light illuminating the sky above the distant thunderstorms. Without knowing it at the time, Robert Franz had observed what became known as "red sprites". Researchers wondered how such a spectacular phenomenon, visible to the naked eye and in our immediate surroundings, could have gone unnoticed for so long. Of course, it had not gone unnoticed: Scientists had not paid attention to eyewitness accounts through the years. (BACKGROUND INFORMATION)

Excerpted with permission from Neuber, T. (2003). On Sprites and Their Exotic Kin. *Science* 300, 2 May, 747.

Although less frequently, the use of surprising information, for example in the form of an analogy, can also be found in more technical or specialized documents as shown below:

Many plants and animals can be considered "smart structures" that can sense and react to their environment. In animals nerve endings are used to sense an environmental effect that could be heat, pressure or light. The signals are then conveyed via nerves to the brain where the signals are then processed and a decision is made on how to react. If a reaction is necessary a signal is sent via another nerve and the animal responds to the environmental effect.
Man-made structures can be made "smart" by duplicating the essential elements of the system that consists of embedded sensors (nerve endings), data links (nerves), a programmed data processor (brain), and actuators (muscles, hormones). Fiber optic sensors offer embedded sensor capability and natural connections to fiber optic data links that can be used in a wide variety of composite materials to act as "nervous systems". (BACKGROUND INFORMATION) These sensors have a series of important advantages over the conventional electronic sensors. (THESIS STATEMENT)

Source: Udd, E. (1996). Fiber Optic Smart Structures. *Proceedings of the IEEE* 84(1), 60. © 1996 IEEE.

5. *A definition.* Sometimes a simple definition serves the purpose of introducing the main idea of the text.

> Bluetooth is a standard developed by a group of electronics manufacturers that allows any sort of electronic equipment – from computers and cell phones to keyboards and headphones – to make its own connections, without wires, cables or any direct action from a user.

Source: Layton J. & C. Franklin. (June 28, 2000) How Bluetooth works. [WWW page]. URL http://computer.howstuffworks.com/bluetooth2.htm. (February 05, 2007). Courtesy of HowStuffWorks. com.

COLLABORATIVE TASKS

3-25 The following groups of sentences constitute a well-written introductory paragraph but they are in the wrong order. In pairs, determine a possible logical order, as there are different possible combinations, bearing in mind that the most general information comes first and that the thesis statement is written last.

Introduction 1

1. In fact computers can do many of the things humans can do but faster and better.
2. They have enormous memories that can store huge amounts of information and do complex mathematical operations in a fraction of the time a mathematician would take.
3. Let us now see some ways in which computers can help people in their everyday life.
4. They can be used to make reservations, keep accounts, control other machines, and even translate a text or play music.
5. Since computers were invented their use has never stopped increasing due to their incredible characteristics.

Introduction 2

1. Particle accelerators can take a particle, such as an electron, speed it up, collide it with an atom and thereby discover its internal parts.
2. However in the second half of the 20th century experiments conducted with "atom smashers", or *particle accelerators*, revealed that the subatomic structure of the atom was much more complex.
3. It was found that the atom was made of smaller pieces called subatomic particles.
4. These particles were basically protons, neutrons and electrons.
5. Hence the importance of these amazing devices which made it possible to know more about the fundamental structure of the matter, the forces holding it together and the origins of the universe.
6. The discovery of the structure of the atom was one of the most remarkable breakthroughs made by physicists during the early 20th century.

3-26 Individually decide which of the following thesis statements are correct. Then check your answers with those of your partner and together specify the subtopics and the associated pattern of organization, if indicated.

> *Example*: Since the birth of the telephone, the world telecommunications network has evolved both in size, slowly at first and then with astonishing speed, and in the technique used to transmit information.
> *Subtopics*: the evolution of the world telecommunication network in size and in the technique used to transmit information
> *Pattern of organization*: Chronological time

1. Smoke signals were an ancient communication system used by Native American Indians and Chinese, who burned damp leaves on a high place to create clouds of dense smoke which could be seen for miles and were used for different purposes.
2. Among the most important ways of communication systems we can distinguish the analogic and digital systems.
3. Videotex has an interactive capability. Teletext simply provides information.
4. Investment priorities in the Third World are often very different from priorities in the industrialized world.
5. Computers have observed an incredible development since the first ENIAC was invented.
6. A whole range of satellites orbit the Earth and are used for a variety of purposes.
7. Process X is a short and simple process as it is basically constituted by three main steps: A, B and C.
8. Alternative means of transportation to the polluting diesel engine would be the hydrogen engine as well as the solar engine.
9. The importance of telecommunication services in the infrastructure of a country.
10. Working under high-pressure conditions for a long period of time may lead to stress, sleep disorders and fatigue.

3-27 The following thesis statements are incomplete. In pairs, finish them by adding possible subtopics.

1. The most important causes of faulty construction in general are...
2. Technological societies are more complex than non-technological societies because...
3. The reduction of noise pollution in big cities is very difficult to achieve owing to...
4. Having a good command of communication skills can help in both...
5. The use of the email has proved to be useful in several ways...
6. The purpose of this report is to...
7. A good word processor must...
8. In comparison to television, radio has some disadvantages, such as...
9. There are different ways of communication or data transmission, namely...

3-28 In pairs write a thesis statement for each one of the following topics. Remember you may specify not only the subtopics but also the pattern of organization:

- Recording information
- Uses of robots
- Artificial Intelligence
- Instilling working habits
- Carbon 14 discovery

CRITICAL THINKING

3-29 The following introductory paragraph has not been adequately written. Say why it is not a good introductory paragraph and suggest ways to improve it.

> Satellites are fundamental tools to modern communication technologies. They were used primarily for military purposes, in navigation and espionage activities. Nowadays they are an essential part of our daily lives. We see and recognize their use in weather reports, television transmission and telephone calls. In many other instances, satellites play a background role that escapes our notice. For example, the most reliable taxi drivers are sometimes using satellite-based Global Positioning System (GPS) to take us to the proper destination. So we can say that what began as an exotic, single purpose device has become a popular, widely-used tool.

3-30 Think of an introductory paragraph in which the thesis statement clearly anticipates the different subtopics and their pattern of organization. What are the advantages and disadvantages of this type of thesis statement as compared to one which does not outline the subtopics?

The body

The paragraphs in an essay other than the introduction and the conclusion are known as *body paragraphs*. Body paragraphs contain most of the information in an essay and can differ in length and number according to the purpose of the document and to the essay pattern you have chosen. The body of an essay is, therefore, the core of a document. The number of paragraphs in the body will be determined by the number of subtopics the writer wants to encompass and by the approach to the topic of the document. It is worthwhile mentioning that a fundamental rule in writing is that whenever you change topic, you start a new paragraph so that you stick to a single topic within the same paragraph. To summarize, the writer's decision on the number of paragraphs and on the order of subtopics will be dictated by:

- the writer's purpose.

- the pattern of essay organization you choose (which could be roughly equated with the *approach* given to the topic).
- the kind of document you have to write.

In the case of highly conventionalized documents, writers should adhere to the pre-established format unless they had some reason to flout the conventions accepted in the community. For example, application letters and research reports are expected to display an approximation to the following structure or format:

Application letters:

Introduction (purpose statement)
One or several paragraphs on *education*
One or several paragraphs on *work experience*
One paragraph on the applicant's *motivation /availability / willingness* etc
Salutation and closing lines

Reports:

Abstract (optional)
Introduction
Method
Results
Discussion
Conclusions

Yet, if there are no conventions governing the format of this document, then the structure of the body is the writer's decision. Most usually, the decision on how information is going to be organized should be made at the outlining stage in a more or less preliminary and perhaps even intuitive way. As writers, you'll have to decide whether the most suitable essay pattern is the one you had already chosen when outlining, or whether you should reconsider the approach and make some changes or adjustments.

INDIVIDUAL TASK

3-31 Read the three introductions below paying special attention to the different thesis statements and, from the information provided in them, write the corresponding body paragraphs for each of them.

> **Introduction I**
> It is obvious that television has become one of the most important inventions of the 20th century. Thanks to technological advances, television has evolved unimaginably, during the past few years, from black-and-white TV to multiple channel services. One of the latest advances has been DTT (digital terrestrial television), which in many

aspects represents a remarkable step further in television technology. If you consider buying a new DTT decoder, there are several points that should be taken into account.

Introduction 2
The rate of road casualties and accidents on the road does not seem to cease with the passing of time. This accounts for both drivers' and car manufacturers' growing concern to enhance vehicle safety systems. Given that automobile safety has become an important marketing tool, we are going to provide an overall picture of current safety systems and delve into their two main types, namely passive and active safety systems.

Introduction 3
Our company has been growing considerably these past few years and is one of the most important in the sector, having a very good reputation in Europe and in the USA. If we want to stay competitive and not be left behind in such a technologically challenging context, the company should employ five engineers for the production department to help the company face this challenge. The purpose of this report is to recommend the most appropriate profile that these five engineers should have. This report will include the criteria used, the discussion of different profile analyses and a final recommendation.

3-32 Read the following outline, which already includes a full introduction and conclusion, and from the information provided develop the different body paragraphs using the pattern of paragraph development you prefer.

I. Introduction: Since oil and its multiple applications were discovered, back in the middle of the 19th century, it has fuelled the modern world. The influence oil has had around the world has no equal but nowadays the two main negative consequences of its widespread use have become readily apparent —its limited nature and its polluting effects; these are the most important drawbacks of non-renewable sources of energy. Thus, humans have realized that we have to look for alternative renewable sources of energy.

II. Main non-renewable sources of energy.
        II.a oil
        II.b natural gas
        II.c coal
        II.d nuclear fission

III. Main renewable sources of energy
        III.a solar energy
        III.b wind energy
        III.c wave and tidal energy
        III.d biomass
        III.e fusion

IV. Conclusion: To sum up, it is clear that we should make a concerted effort to reduce our dependence on non-renewable sources, as the day will soon arrive when the cost of obtaining oil and natural gas (the two energies that make up fifty percent of our current supplies) will be higher than their value. Conversion to a completely renewable and sustainable energy economy will certainly bring about changes in lifestyle because sustainable energies will not be able to sustain present rich world levels of energy and living standards. But this is not the only challenge lying before us. Humankind—and engineers in particular—is now challenged with the limits of technology.

The conclusion

The conclusion is the last paragraph of the essay and signals the reader that the essay has come to an end and that no more new ideas will be tackled. It can have two parts: *the summary or restatement of the main points sketched out in the thesis statement* and/or *a final comment*. Ending the essay with a summary or restatement of the main points is useful inasmuch as it helps the reader remember the main points discussed in the essay and, at the same time, contributes to the thorough understanding of the topic. More often than not, the summary is followed by a final comment, which can take many different forms:

- a recommendation
- a prediction
- a solution
- a personal comment
- a question
- a quotation

It is interesting to notice how, if compared with the introduction, this part of the essay organizes the information in the opposite direction, that is, from specific to general. The writer refers back to the information in the thesis statement, restating it, and ends with a final general comment.

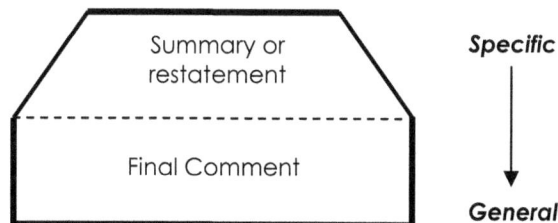

Fig. 3.6

Note that the use of a transitional marker such as *In conclusion, To sum up, Thus,* etc.[9] contributes to signalling that the reader has reached the conclusion or the end of the essay. Also remember that the conclusion should be neither too short (one or two sentences)

---

[9] Some of these can be found in the connecting expressions table in Section 3.5.

nor too long (several paragraphs) and that you should add new ideas to the body paragraphs and not here at the end when closing the essay.

The following example of a conclusion belongs to the essay on the advantages of mobile phones.

> In conclusion, it is reasonable to say that the great impact that mobile phones are having on our society is mainly due to their advantages. Mobile phones have made it possible not only to call somebody from almost anywhere thanks to their wireless technology but also to obtain information from the Internet or have fun playing games. Yet, in spite of the advantages, we cannot forget that, as with most things, mobile phones have also some disadvantages that cannot be underestimated. (Personal comment)

In this sample conclusion, firstly we have a summary of the main advantages of mobile phones and then the personal remark that the advantages cannot be thought of independently from the disadvantages. The same conclusion, nevertheless, can be finished with different final comments according to the final thought with which we want to leave the reader. For example, if we want to finish with a *prediction* then we can write:

> Because technological development is improving day by day we might dare to say that mobile phone applications will not stop increasing as long as operators assume that the new applications are beneficial to their pockets. (Prediction)

But if we want to end up with a question, then we can write:

> So the question is: with all the applications offered by the 3G mobile phones, when will technical experts be able to develop a device which substantially improves the 3G mobile capacity of serving the triple purpose of communicating, informing and entertaining? (Question)

Some conclusions will neither follow the patterns of information organization presented above nor have the same position or the same length. Certain documents such as administrative reports, for example, have the conclusion after the introduction or the method, not at the end of the document where it is normally located. Because readers of this type of document are especially interested in this part of the document, it was probably thought most convenient to place it earlier in the document. Likewise, the concluding information in this type of reports is not based on a summary or restatement of the main points but rather on putting forward the main results achieved. The results are normally discussed and considered for further action. These reports, at the same time, may have a conclusion longer than a paragraph, even occupying a full section. Thus, conclusions as well as other parts of a document will have different forms depending on the specific type of document they belong to.

INDIVIDUAL TASKS

3-33 The following is a well-written conclusion. First, try to identify the restatement of the main points and the final comment and then say what type of final comment has been used.

> We can conclude by saying that pollution is becoming a serious and wide-ranging problem as it is affecting not only the air we breathe but the water we drink and the food we eat. Besides, the levels of pollution are increasing at a fast, alarming pace, thus posing our planet in real danger. So governments should become more concerned with this problem and should promulgate strict laws against it before it is too late to save our planet.

3-34 The following conclusions are incomplete because they only have a summary or restatement of the main points. Finish them by adding a suitable final comment.

> **Conclusion 1**
> In conclusion, technological advances have brought many advantages to modern life such as a reduction of the world's distances, an improved medical assistance and a better productivity thanks to automation. But these advances have also brought some disadvantages, which have resulted in a degradation of human inter-relationship, a more polluted environment and a growing difference between rich and poor countries...
> ....
> ....

> **Conclusion 2**
> To sum up, television has achieved to entertain both adults and children. It has become an essential medium of entertainment in our society. In particular, children are so fond of it that they can spend hours in front of a TV screen watching all sorts of programs. This indiscriminate use of television has demonstrated to have a negative effect on them since it reduces the time they should spend on other necessary activities such as practicing sports, reading and studying..
> ....
> ....

3-35 Write a simple, complete conclusion for the introductions in task 3-25. Begin by identifying the main points or thesis statement in the introductions. Once you have identified them, proceed with writing the conclusions. Do not forget to use a transitional signal.

3-36 The text below is a first draft of an essay on energy saving written by an engineering student. Read it and then in pairs decide in which ways it can be improved, focusing on the aspects related to the structure of the essay. Grammatical accuracy is not yet a matter of concern at this stage even though you can obviously correct as many mistakes as you wish.

Nowadays, in developed western countries life is easier and more comfortable than ever before. If we want to see in darkness, go up ten floors, talk to somebody that lives in a far place, listen to our favourite rock band ... we only need to press a simple key. Furthermore, you can go somewhere by car or plane in a very short time. But not all are advantages. On the other hand, most of the devices mentioned before work with electricity, a source of energy that pollutes a lot. And what's more, another problem is those energies that will be used up in the future, such as petrol. For all these reasons, in my opinion, we should do something and start using other alternative energies that do not pollute the environment so much. For example the Balear government will enforce a law that will demand all new buildings be built with solar energy and using wind power, too. Also, if we want to keep using electricity, measures should be taken at every private home; for instance, always turn off the light and unplug all appliances when we no longer use them or need them. With regard to vehicles, one novelty about them is their use with solar energy or hydrogen. In summary, with measures taken by the State and as far as we are concerned and make an effort to be more respectful for our environment, problems will probably disappear gradually.

CRITICAL THINKING

3-37 Taking into account the outline provided, read the following conclusion and decide whether it is a good conclusion or not and say why.

Topic: Laser applications
Outline:
   1. *Introduction*
   2. *General description*
      2.1 Origin
      2.2 Definition
      2.3 Main characteristics
   3. *Applications*
Communications
Semiconductor devices
Reading mechanisms
Medicine
Surgery
Eye treatment
            3.5 Industry
               3.5.1 Jewellery
               3.5.2 Printing
   4. Conclusion: In brief, the laser has proved to be one of the most useful, modern inventions thanks to its associated properties which have found a wide range of applications in different fields, namely in communications, medicine and industry. To these applications we should add the domestic one in devices such as alarm clocks, DVD players and keys. So we can say that the use of the laser has never stopped increasing since its invention and it will not stop until anything better can replace it.

3-38 How would you react in front of a technical text which begins with an introduction that includes the writer's opinion?

3.7 Developing essay patterns

As we saw in the Developing paragraph patterns section, different topics call for different patterns of organization. The same could be said about essay patterns, which are essentially the same patterns but operating at a more global level. As with paragraph patterns, the subtopics should be organized using a consistent pattern that arises naturally from the thesis statement and that derives from the writer's purpose. The most common patterns in essay organization in technical and scientific documents are basically the same as the paragraph patterns studied before:

- chronological or sequential order
- cause and effect
- comparison and contrast / analogy
- specific to general
- general to specific
- problem and solution
- listing / order of importance

Below are several examples of outlines which reflect how the issue on mobile phones can be approached from different perspectives following different patterns of organization.

Outline based on a *comparison-and-contrast pattern* combined with a (secondary) *listing* one:

Subpattern A: Advantages: advantage 1+ advantage 2+ advantage 3
Disadvantages: disadvantage 1+ disadvantage 2+ disadvantage 3

    I. Introduction
    II. Advantages:
        A. Reachability
        B. Mobility (wireless)
        C. Easy to use and to carry
        D. Information and entertainment service
    III. Disadvantages:
        A. Technical limitations
        B. Social implications
        C. Environmental issues and health risk
    IV. Conclusion

Subpattern B: Advantage 1+ disadvantage 1
                   Advantage 2+ disadvantage 2
                   Advantage 3+ disadvantage 3

    I.   Introduction
    II. Advantage 1+ disadvantage 1
       A.   Reachability, mobility, portability
       B.   Coverage, short-life battery, privacy
    III. Advantage 2+ disadvantage 2
       A.   Information service
       B.   B.short-life battery, new SMS language, addiction
    IV. Advantage 3+ disadvantage 3
       A.   Entertainment service
       B.   Lack of face-to-face communication, addiction
    V.   Conclusion

Outline based on a *specific-to general essay pattern*, reflecting the deductive method of reasoning:

    I.   Introduction
    II. Applications
       A.   First generation applications:
          A.1. Telephone calls
          A.2. SMS
          A.3. Phone book, agenda, calendar
          A.4. Message box
       B. Second-generation applications:
          B.1. Voice mailbox
          B.2. E-mail service
          B.3. SMS
       C. Third-generation
          C.1. MP3
          C.2. Internet access
          C.3. Pictures and video-conferencing
    III. The present and future prospects
       A. Mobile phones as an example of technology changing society
       B. Massive use and explosion of mobile phones
       C. The future of mobile technology
    VI. Conclusion

Outline based on a *general-to-specific essay pattern*, reflecting the inductive way of reasoning:

    I.   Introduction
    II. Mobile phones nowadays
       A. Explosion of mobile phones
       B. Multi-functionality

III: Technical features of mobile phones:
    A. Physical characteristics: size, measures, weight, ...
    B. Components
IV. Conclusion

Outline based on a *problem-solution essay pattern*:

I. Introduction
II. Past problems without mobile phones
    A. Difficult reachability
    B. Lack of mobility in fixed phones
    C. Lack of portability in fixed phones
III. Present problems with mobile phones:
    A. Technical limitations
    B. Social implications
    C. Environmental issues
IV. Solutions to present problems
    A. Further research to improve technology
    B. Actual liberalization of the telecommunications sector
    C. Education on how to use mobiles appropriately
    D. Legislation
V. Conclusion

Finally, other points to take into consideration in relation to essay patterns when organizing information in a technical document are:

- Not all documents must necessarily be entirely based on the essay patterns above. Sometimes they may have full sections or chunks of text following no specific pattern.
- As with paragraph patterns, it is quite common to find an essay with two or even several patterns embedded.
- It is usually worthwhile to pay special attention to connecting devices. From a writer's point of view, you are bound to make use of connecting devices that reflect the essay pattern you are following. For example, if you have decided to base your writing on a cause-and-effect pattern, you will very likely resort to connective devices that signal causes or effects (e.g. *for this reason, since, therefore*).

INDIVIDUAL TASKS

3-39 Reread the sample essay on mobile phones (page 132) and identify the essay pattern used.

3-40 Imagine you have to write an essay about either of these two topics **a)** *New applications of electronic systems or* **b**) *Vehicle safety systems.* They are very broad titles so that different writers can deal with them from completely different approaches.

- Write up two outlines to the same essay, one based on a listing essay pattern and one on a cause-and-effect essay pattern.
- Now choose one of your outlines and write a full introduction including the thesis statement and the topic sentences for every body paragraph.

COLLABORATIVE TASKS

3-41 In pairs, skim the outlines below and say which essay patterns they seem to be based on. Some of them could be greatly improved. Can you spot the weaknesses and suggest any improvement? In the cases where the essay pattern is not clear, consider improving the outline first taking into account that there may be various possibilities of organizing information in a logical way.

Outline 1. (The Internet)

```
    I.   Introduction
    II.  Communication
            A. Contact everybody
            B. Face-to-face communication and human relationships are impoverished
    III. Information
            A. all kinds of information
            B. illegal webs, illegal uses of information
    IV.  Conclusion
```

Outline 2. (The Internet)

```
    I.    Introduction.
    II.   Easy communication and information
    III.  Viruses, hackers
    IV.   Games
    V.    Quick communication and information
    VI.   Inadequate, illegal information
    VI.   Music
    VII.  Quantity of information
    VIII. Free music (musicians' copyright)
    IX.   Conclusion
```

Outline 3. (The Internet)

I.   Introduction
II.   First purposes and goals of the net
III. Slow growth along with computer technologies and networking
IV.   Improvement over existing sharing networks (like CompuServe)
V.   Remarkable growth with the availability of PCs with Windows95 running navigators.
VI.   Fast expansion: it has no frontiers
VII. Uncontrolled content: starting to cause problems
VIII. Conclusion

Outline 4. (The Internet)

I.   Introduction
II.   Pornography without censorship
III. Pornography sites should be difficult to access
IV.   Loss of concentration at work
V.   Employers can register the staff connections
VI.   Marginalization of those without the Internet
VII. Ease of access in cyber cafés
III. Conclusion

Outline 5. (The Internet)

I.   Introduction
II.   The Net as a great source of information
    A. computers link us to a useful tool
    B. all kind of information
III. Fast way of being connected to the world
    A. buying, learning, travelling… through the net
    B. a 'space' without social status
IV.   The Internet: a waste of time and money
    A. a bad hobby at a high cost of money and time
    B. another way to break the law
V.   Conclusion

3.8 Providing inter-paragraph coherence

As defined in the Providing intra-paragraph coherence section, *inter-paragraph coherence* envisions the essay from a global perspective (different paragraphs that make up the text) and refers to how coherence can be achieved *between* paragraphs. As already indicated, there is some parallelism between the structure of a paragraph and that of an essay; similarly, there is some parallelism between *intra- and inter-paragraph* coherence. As with intra-paragraph coherence, there are three main ways to make an essay *cohere;* that is, to organize and connect paragraphs in such a way that the reader perceives that the paragraphs are integrated and form *one* text:

- Shared knowledge
- Thematic unity
- Transition signals

Shared knowledge

When you are writing to a peer audience, as a writer you could in fact decide to write a document implicitly. That is, you could write a document with no section headings or with no lexical signals that explicitly tell the reader the function and connection of one or several paragraphs in relation to the previous or subsequent one, and nevertheless write it with coherence at the same time. In this case, what makes the text cohere and appear as *one* text is the knowledge of the content shared between writer and reader. The burden of interpreting the message is then passed onto readers, who will have to infer the meaning with the help of their specialized knowledge. Such implicit texts that reflect that the writer does not guide the reader are known as *writer-oriented* texts.

Below is an example of a two-paragraph text that is coherent for an expert in the topic. Expert readers will realize there is a connection between these two paragraphs. They will understand that the first paragraph acts as an introduction where the writer expounds the widely-accepted belief that the market is a better regulator than the state. The second paragraph corroborates this assumption because regulators' estimates are never good enough. The examples of the two UK utility companies show that government regulation does not always attain its goals. This relationship is only supported by two examples but could be made more explicit by introducing the second paragraph with a connecting expression like *to illustrate this point*. On the other hand, for non-expert readers not sharing knowledge on such specialized topic the text may sound double Dutch and may even appear as two disconnected and unrelated paragraphs:

UK utilities

As Soviet Russia discovered, government efforts to allocate investment often end in tears. Today the market is generally accepted as a superior mechanism for establishing the cost and allocation of capital.
Most UK utilities can make only a fair return on their regulatory asset base. Regulators' estimates of this vary hugely. Ofgem has produced pre-tax nominal WACC estimates as low as 8 percent and on one occasion Oftel drifted to a delirious 17 percent.

Adapted from: The Lex Column. (November 15, 2004). Mud-caked. UK utilities. *Financial Times*, 14.

Thematic unity

Essentially, thematic unity here refers to thematic unity *across* the essay, *between* paragraphs, which means that paragraphs are related in topic. At the global level we are now, thematic unity does not pose many problems to technical writers as many technical documents have headings and subheadings which give every paragraph or section a function within the whole document. Taking things to an extreme, even the outline for writers and the Contents Table for readers could be regarded as elements giving thematic unity, preventing them from drifting off the track. With documents or passages without sections or headings, coherence can be achieved:

a) when the thesis statement unfolds so that no off-topic jump appears unexpectedly in a given paragraph
b) when every paragraph deals with the subtopics following the same order as hinging from the thesis statement.
c) when there is structural parallelism across paragraphs; that is, when information is repeated in a similar order.
d) when a paragraph pattern is used consistently.

Notice how the first example has thematic unity since the body paragraphs are related in topic and display the subtopics announced in the thesis statement, whereas the second example overlooks this unity by adding an unexpected subtopic and by reversing the order of subtopics.

Introduction	***General statements*** + Friction should be avoided because it can be destructive and wasteful on many occasions. ***(Thesis statement)***
Body paragraph 1	Friction can destroy and damage components in different ways.
Body paragraph 2	Because it is also wasteful, friction can cause serious and expensive breakdowns.

Introduction	***General statements*** + Friction should be avoided because it can be destructive and wasteful on many occasions. ***(Thesis statement)***
Body paragraph 1	Because it is wasteful, friction can cause serious and expensive breakdowns.
Body paragraph 2	~~Lack of regular maintenance can lead to friction.~~ *(off-topic!)*
Body paragraph 3	Friction can destroy and damage components in different ways.

The third way to achieve thematic unity and coherence is through parallelism across paragraphs. When readers find two paragraphs with the same internal parallelism, they will immediately interpret this parallel structure as an integrative signal. The strategic repetition of the same or similar order of information and of the same syntactic, lexical or grammatical structures constitutes parallelism that functions at a global level. For example, the structural parallelism across the second and third paragraphs below is instructing readers that they form one text. The repetition of the structure "definition + classification + examples of applications" indicates that the paragraphs are performing a similar function, thus acting as format markers. On top of that, the parallelism in the list-like structure also ties the text together and so highlights coherence.

> Metals are probably the materials most closely associated with the engineering profession. The major mechanical properties of metals are stress versus strain, hardness, impact strength, fracture toughness, fatigue, and creep. Metals can be divided into ferrous and non-ferrous metals.
>
> Ferrous metals are metals containing iron. These alloys fall into two broad categories based on the amount of carbon in the alloy composition: i) carbon and alloy steels, and ii) cast iron. They are used in the majority of metallic applications in current engineering design like ball bearings or metal sheet formed into automobile bodies.
>
> Non-ferrous metals are metals that do not contain iron as the major constituent. They fall into ten main categories based on the amount of the alloying element: i) aluminium alloys, ii) magnesium alloys, iii) titanium alloys, iv) copper alloys, v) nickel alloys, vi) zinc alloys, vii) lead alloys, viii) rapidly solidified alloys, ix) refractory metals, and x) precious metals. They are used in a wide range of technological applications like advanced aircraft, consumer products, radiators, automobiles and hardware, coatings for catalytic converters, or gold circuitry in the electronics industry.

Transition signals

Transition signals are used to tie up ideas in a text, acting like bridges between paragraphs. They can be long or short, simple or more complex because they can be made up of only one word, a connecting expression, a phrase, a sentence or even a paragraph. They are usually placed at or near the beginning of the paragraph, even though transition sentences in particular can appear at the end of a paragraph in order to make a link with the topic of the next paragraph whereas in longer texts, we can even find a transition paragraph. Whatever their length and composition, they always have the same basic purpose: they look back at what has been said and at the same time look forward to what will be dealt with.

Writers may choose the degree of explicitness they prefer or the one they regard as more appropriate to their audience. Making an efficient use of transition signals throughout an essay has proved to be beneficial given that they provide coherence. More specifically, transition signals can be said to give coherence for at least three reasons:

- they are ambiguity-reducing devices that guide and orient the reader through the whole text. In this way, the reader is not left alone to figure out the underlying relationships or connection between one paragraph and the other.
- they help carefully distribute known and new information, thereby enhancing coherence and letting ideas flow smoothly from one paragraph to another.
- they usually contain repeated key words and pronouns or synonyms that substitute them, which enhances coherence and facilitates comprehension.

Finally, texts which deploy an efficient use of such transition signals are called '*reader-oriented*' texts. Let us now tackle transition signals in more detail.

Short transition signals

These are mainly connecting expressions which typically work at the global level, many of which were already mentioned in the Developing paragraph patterns section (page 97) and in the Providing intra-paragraph coherence section (page 113). These transition signals usually reflect the essay pattern or make it apparent. The most common types are:

- *one-word transitions* inserted in the topic sentence
- *sentence connectors* preceding the topic sentence
- *phrases and subordinated clauses* also preceding the topic sentence

Below you can find some examples:

> The **first** factor to consider is X (transition word)
> **Next**, + topic sentence (transition sentence connector)
> **Nevertheless**, + topic sentence (transition sentence connector)
> **In addition to this**, X is another factor worth considering (transition phrase)
> **Apart from this**, + topic sentence (transition phrase)
> **As regards X**, + topic sentence (transition phrase)
> **In spite of this**, + topic sentence (transition phrase)
> **After having seen how 'X' works**, + topic sentence (non-finite transition clause)
> **Although X has been considered crucial**, + topic sentence (subordinated transition clause)

Long transition signals

Long transition signals can be found either at the beginning of a paragraph—preceding the topic sentence—or at the end. Their function is the same as that of short transitions; that is, they mostly link what has been said in the preceding paragraph(s) with what is going to be written onwards. Sometimes short and long transitions can be used in combination. For example, note how the idea of contrast can be conveyed by means of one word connector, a transition sentence, or both.

Body paragraph 1	Hybrid engines are a good alternative to combustion engines. *(+ supporting sentences)*
Body paragraph 2, with transition signals	**OPTION 1 However**, their manufacturing cost is much higher than that of combustion engines. *(+ supporting sentences)* **OPTION 2 Hybrid engines still have some limitations.** Their manufacturing cost is much higher than that of combustion engines. *(+ supporting sentences)* **OPTION 3 However, hybrid engines still have some limitations.** Their manufacturing cost is much higher than that of combustion engines. *(+ supporting sentences)*

Long transition signals could be further subdivided into three main types:

- *Transition sentences connecting two paragraphs.* They are made up of a sentence which can appear in combination with short transitions like a connector. Just like short transitions, they usually precede the topic sentence but they can also be found at the end of the previous paragraph.
- *Transition sentences that go over one paragraph.* They are like topic sentences that make the transition between one section and another or between a block of several paragraphs and another one.
- *Transition paragraphs.* They are paragraphs which connect the preceding paragraphs with the forthcoming ones. They are usually found in long documents.

Below are some examples of transition sentences that appear at the beginning of the paragraph and connect two paragraphs:

It is now worth mentioning another factor, which is + topic sentence
There is another remarkable aspect to bear in mind apart from the 'x' and 'y' factors mentioned above. + topic sentence
Additionally, another point should be made. + topic sentence

The following example contains two transition sentences which appear at the end of a paragraph and which are marked in grey. Notice that they are the last sentences of a paragraph and make a link to the topic sentence of the following paragraph. In this same example, we can also find a transition sentence that goes over more than one paragraph. As you can see, the transition sentence in body paragraph 3 can be regarded as a transition signal between the first block (advantages in body paragraphs 1 and 2) and the second block on disadvantages (body paragraphs 3, 4, 5 and 6). This transition sentence appears white on black.

Introduction	*(general statements)* + Even though mobile phones are an indispensable gadget nowadays, not all is advantageous about them *(thesis statement).*
Body paragraph 1	Needless to say, thanks to mobile phones communication is easier and more comfortable than ever before. *(+ supporting sentences).* **Ease and comfort are not the only advantages.** *(Transition sentence)*
Body paragraph 2	Immediacy and availability are **also** all-important factors we can enjoy thanks to cellular phones. *(+ supporting sentences).* **As one could imagine, we are paying a price to all these benefits**. *(Transition sentence)*
Body paragraph 3	**Nevertheless, mobile phones have some drawbacks that cannot be overlooked.** **First** is their need to be recharged *(+ supporting sentences): transition sentence + topic sentence*
Body paragraph 4	**Moreover**, mobile phones are still more expensive than traditional phones, at least for the time being. *(+ supporting sentences)*
Body paragraph 5	**In addition to their dependence on batteries and their price**, mobile phones have entailed another incontestable disadvantage, namely loss of privacy. *(+ supporting sentences)*
Body paragraph 6	**Last but not least** is the most worrying and thorny issue of whether exposure to microwaves represents a risk for human health over long periods. *(+ supporting sentences)*
Conclusion	**To conclude, ...**

Finally, an example of a transition paragraph can be read in the reduced version of the essay below:

Introduction	The decrease in the economic activity in most markets for the last few years resulted in several companies cutting the expenditure on research and development (R&D), thus collaborating to a certain economic stagnation in several markets. In contrast, the mobile industry showed a stable growth and was constantly investing in the development of new or better products. As a matter of fact, the role of the mobile phone in our present society is remarkable: the mobile phone has invaded every home and reached every family, bringing about different ways of communicating and socializing with

	people. The social influence of the mobile phone calls for a deeper study of the advantages and disadvantages of mobile phones.
Body paragraph 1	Obvious advantages can be found in x, y, and z.(...)
Body paragraph 2	Other advantages depend on the personalization of the mobile phone and on future developments. (...)
Body paragraph 3, **Transition paragraph**	**This all sounds great and it must be admitted that the mobile phone has become part of our life. This second part of the essay will show that it is not only happiness what mobile phones have brought us. The negative aspects that will come up are on the one hand related to their impact on social life, and on the other related to the technical side effects of a widespread mobile use**.
Body paragraph 4	Due to the great use of mobile phones, there is a growing feeling that mobiles are disturbing social life. (...)
Body paragraph 5	The technical disadvantages rely on the interference of the channels used by the mobile and other equipment, (....)
Conclusion	To conclude, the mobile phone can be stated to be an essential part of modern life. The advantages make the mobile phone an all-round device which provides people with the opportunity of having everything they want. But to make the mobile phone an even greater success, the companies involved as well as the governments should show some concern for the disadvantages and seriously consider taking action so that the negative side effects of mobile phones are no longer a matter of concern.

Use of key words, pronouns and synonyms within transition signals

Other ways of providing the text with coherence are: (i) to repeat some key words present in the thesis statement or in previous topic sentences, (ii) to use synonyms of key words, and (iii) to incorporate pronouns like *these, this, the latter* referring back to already mentioned concepts. These words can be inserted in transition signals, topic sentences or conclusion paragraphs. Look at the example below to see how the repetition of the same key word used in the thesis statement can be a good way of making a text cohere:

Introduction	*General statements* + Even though mobile phones are an indispensable gadget nowadays, not all is **advantageous** about them *(thesis statement).*
Body paragraph 1	A remarkable **advantage** of mobile phones is that, thanks to them, communication is easier and more comfortable than ever before. *(+ supporting sentences).*

Sometimes, however, it is preferable to avoid repetitious beginnings of paragraphs. If, for example, you have already repeated one or two key words in the transition or topic sentence of the first paragraph, it may be useful to look for a synonym—provided readers understand you are referring to the same concept—in the following paragraph.

Introduction	General statements + Even though **mobile phones** are an indispensable gadget nowadays, not all is **advantageous** about them *(thesis statement).*
Body paragraph 1	A remarkable **advantage** of mobile phones is that, thanks to them, communication is easier and more comfortable than ever before. *(+ supporting sentences).*
Body paragraph 2	Immediacy and availability are also **all-important factors** we can enjoy thanks to **cellular phones**. *(+ supporting sentences).* As one could imagine, we are paying a price to all **these benefits.**

As you can see, the key word *advantageous* in the thesis statement is repeated in *advantage*. Similarly, this key word (*advantageous*) is rephrased by means of two synonyms *all-important factors* and *benefits*, and *mobile phones* is also rephrased as *cellular phones*. The use of the cohesive pronoun *these* to refer back to the main ideas put forward in the first and the second body paragraphs also helps the text cohere.

In brief, there are four main ways to link paragraphs explicitly by means of a marker:

- With a short transition signal (e.g. first, firstly, in the first place, another, apart from +ing)
- With a transition sentence at the end of a paragraph or at the beginning of another
- With a transition sentence that goes over more than one paragraph
- With a transition paragraph

In all of them, we can repeat the key words present in the thesis statement or use synonyms and pronouns. By now you may be wondering 'which transition signal should I use?' Well, it's up to you, as all are good enough. In fact, you do not need to stick to just one type of transition within the same document: you can combine different types, thus adding variety and richness to your writing.

COLLABORATIVE TASKS

3-42 For the three introductory paragraphs below, identify the thesis statement and from the information provided in it, write the topic sentences for every body paragraph. Remember to use different types of transition signals to mark explicitly the transition between paragraphs.

Introduction 1
Computers and other equipment have a very short life today. In other words, when two years have passed since they were first bought, they soon become 'e-waste'. This is the term used to refer to electronic waste, all the useless objects or elements that come from an old-fashioned electronic device and which some people dispose of carelessly. Since there is an ostensive lack of awareness about this issue, e-waste is hardly recycled and has become a serious polluting agent.

Introduction 2
Nowadays it is extremely difficult to find a person who assures is not concerned about the damage human beings are causing to our planet. We have in our hands some ways to diminish environmental damage, from recycling to promoting technologies like tidal power generators. Lately we have heard of a new kind of house called 'intelligent house', an alternative to the conventional dwelling. Intelligent houses incorporate some of the latest technological advances and environmental-friendly facilities not only to improve our comfort but also to contribute to our planet's sustainability.

Introduction 3
When designing a race car, engineers' main objective is to make it as fast as possible on certain conditions while observing several technical specifications. From this point of view, it is worthwhile comparing two vehicles with very different conceptions. On the one hand, F1 cars are designed to be fast on various tracks where a high average speed is achieved. On the other, WRC (World Rally Championship) cars have a road car origin and are designed to be fast on different terrains such as mud or water. This essay will analyze their main technical aspects—engines, suspension, brakes, and transmission.

3-43 Rewrite the essay below in order to improve it concentrating on the following two aspects:

- paragraph organization and topic sentences
- *inter-* and *intra-* paragraph coherence (connecting devices, unity, parallel structures, use of key words and pronouns).

Modern Materials Needs

Materials are probably more deep-seated in our culture than most of us realize. Transportation, housing, clothing, communication, recreation, and food production—virtually every segment of our everyday lives is influenced to one degree or another by them. The development of many technologies that make our existence so comfortable has been intimately associated with the accessibility of suitable materials. An advancement in the understanding of a material type is often the forerunner to the stepwise progression of a technology, as in the case of sophisticated electronic devices and automobiles. Even though tens of thousands of materials have evolved with rather specialized characteristics, our modern society needs increasingly more specialized new materials, most of which concern our planet's sustainability.

Energy is one current concern. There is a recognized need to find new economical sources of energy and, in addition, to use the present resources more efficiently. Materials can undoubtedly play a significant role in the developments. Solar cells used to convert solar energy into electrical energy employ some rather complex and expensive materials. To ensure a viable technology, materials that are highly efficient in this conversion process must be developed so that they are even less costly.

Nuclear energy also holds some promise, but the solutions to the many existing problems will necessarily involve materials, from fuels to containment structures to facilities for the disposal of radioactive waste.

Furthermore, research on pollution-free methods should be carried out following two directions. As it is known, environmental quality depends on our ability to control air and water pollution, so the more respectful and efficient the materials involved in many pollution control techniques, the environment will be better. Materials processing and refinement methods also need to be improved so that they produce less environmental degradation, that is, less pollution and less despoilage of the landscape from the mining of raw materials.

Significant quantities of energy are involved in transportation. In order to enhance fuel efficiency, we can both reduce the weight of transportation vehicles (automobiles, aircraft, trains, etc) and increase engine operating temperatures. New high-strength, low-density structural materials remain to be developed and we also need to develop materials that have higher-temperature capabilities for use in engine components.

Many materials that we use are derived from resources that are non-renewable; that is, not capable of being regenerated. These include polymers, from which the prime raw material is oil, and some metals. Non-renewable resources are gradually becoming depleted and it is crucial to discover additional reserves and develop new materials having comparable properties and less environmental impact. The alternative of developing new materials with comparable properties and less environmental impact is a major challenge for the materials scientist and engineer. Therefore, a significant concern points to the need of finding not only new materials as efficient as, or more efficient than, the current materials, but also of finding materials based on renewable resources. In spite of all the tremendous progress that has been made in the understanding and development of materials within the past few years, there remain technological challenges requiring even more sophisticated and specialized materials that meet the needs of our modern and complex society. To make engineers' work more difficult, sophisticated and specialized materials should cater for the current situation of our planet and the depletion of non-renewable resources should also be catered. New materials need being developed, with the caveat that new materials and their processing methods ensure the Earth's sustainability. We hope materials engineers will succeed in solving the problems we are faced with, as this would certainly be one important step in the technological progress that contributes to a better world.

Source: *Materials Science and Engineering. An Introduction* by W. D. Callister. Copyright © John Wiley & Sons Inc. 1994, 2-5. Reprinted with permission of John Wiley & Sons, Inc.

INDIVIDUAL TASKS

3-44 Working as an engineer, imagine you have to write an essay on one of the topics below. Notice that these issues are very broad, so you will have to narrow them down and adapt the essay to every writing situation, as described by your teacher. You can also write on any other topic you know well or you like related to the technical or scientific world. When developing the text bear in mind everything you have been taught about essay writing.

Alternative sources of energy	Pollution-free environment
Present and future of telecommunications	Global warming
Bio-engineering	Vehicle safety systems
The Internet	Domotics
Face-to-face communication	New materials and their applications.
Ecological cars	Nuclear fission v. fusion

CRITICAL THINKING

3-45 Studies suggest that different writers from different language cultures use inter-coherence expressions differently. English writers writing in English seem to be more explicit and reader-oriented than some Romance writers writing in English. In view of these different writing practices, which group do you align yourself with? Can you provide a short list of the inter-coherence expressions you usually use?

3.9 Incorporating visual aids

We have already emphasized the importance of a good document design for clarity and readability. The organization of information into relatively small and balanced paragraphs allows the reader to better understand the overall organization of a text. This chunking of information, which also includes an adequate use of headings, white spaces and section dividers, together with different ordering techniques such as typeface and type size, boldfacing, position and spacing after each heading will allow readers to visualize the order of information and the writer's priorities. For this reason, it is important to choose the most adequate techniques for each occasion and be consistent all through the document. On the other hand, in order to present information not only adequately but also effectively and persuasively, sometimes you need to include visual aids. These allow you to provide a lot of information in a clear and concise way and they constitute a visual appeal to the reader, catching the reader's attention and breaking up the monotony of a text. In addition, they give support to your argument and help convince the reader of the validity of the findings. All these benefits favour the technical writers' preference to include visual aids in their texts.

Visual aids can be used in a variety of documents with the purpose of strengthening the message you intend to transmit. Therefore, in order to be communicatively effective, graphics should follow the criteria below:

- be a complement of the text. Graphics may add information, summarize the text, help explain content, confirm information in the text, etc. in a more catchy and unforgettable way than the text alone, which will allow you to communicate more effectively.
- be informative without being redundant.
- be located immediately following the text referring to the graphic, and be linked with the text by means of an appropriate textual reference, for example: *The data in Table 1 show* or *(see Figure 1)*.
- have the format that best displays the intended data and be correctly labelled in order to enhance information and to help readers visualize data.
- be readable, keep an adequate size and be neatly presented.

Types of visual aids

There are many different types of visual aids with which you can improve your document. In order to choose the most appropriate, you have to decide on their strengths and weaknesses and choose the one that better suits your purpose. Nowadays computers allow you to generate many types of graphics easily and to display the same data in different formats, which may ease the task of selecting the most appropriate graphic for each occasion. Some of the most common types of visual aids used in scientific and technical writing are described below.

Tables

Tables are convenient for displaying a large amount of numerical data in a small space and for presenting individual data items precisely. They organize information into rows and columns by category and type, eliminating repetition of words and simplifying access to individual data values, thus making the text more readable. However, although they are useful for providing item-to-item comparisons, they don't show trends or direction in the data. Therefore, for this purpose a more visual graphic should be chosen. Below you can see an example of a table displaying some of these common characteristics.

Table 2. Energy Consumption and Carbon Dioxide Emissions in North Central Europe, 2001

Country	Total Energy Consumption (quadrillion Btu, 2001)	Oil (thousand barrels per day, 2001)	Natural Gas (billion cubic feet)	Coal (million short tons)	Electricity (billion kilowatthours)	Energy-Related CO_2 Emissions (million metric tons of carbon, 2001)
Poland	3.54	424	489	150.7	118.8	78.6
Czech Republic	1.53	176	349	67.8	55.6	29.0
Slovak Republic	0.83	82	280	10.24	24.4	10.8
Hungary	1.09	149	472	17.54	35.1	15.2
Total	6.99	831	1,590	246.3	234	133.6

Table 3.2

Source: Energy Information Administration. (2003). North Central Europe (Table 2: Energy Consumption and Carbon Dioxide Emissions in North Central Europe, 2001) [WWW page]. URL http://www.eia.doe.gov/emeu/cabs/poland.html

Line graphs

Whereas tables present discrete data items, line graphs show data continuity, direction and trends over time. Once all the data are gathered, they should be plotted and linked by means of a line in order to show the upward and downward movements of data. Line graphs work well for comparing different sets of values (for example, figure 3.7); nevertheless, no more than three or four lines should be traced in the same graph because it may be difficult to read and may confuse readers. Also, although these graphs may be useful for revealing relationships between different types of data, they do not describe well the relationship between too many lines, the intersection between them and the significance of those points that fall off the line, so look for another solution if you intend to emphasize any of these aspects.

Production of commercial energy 1980-2002

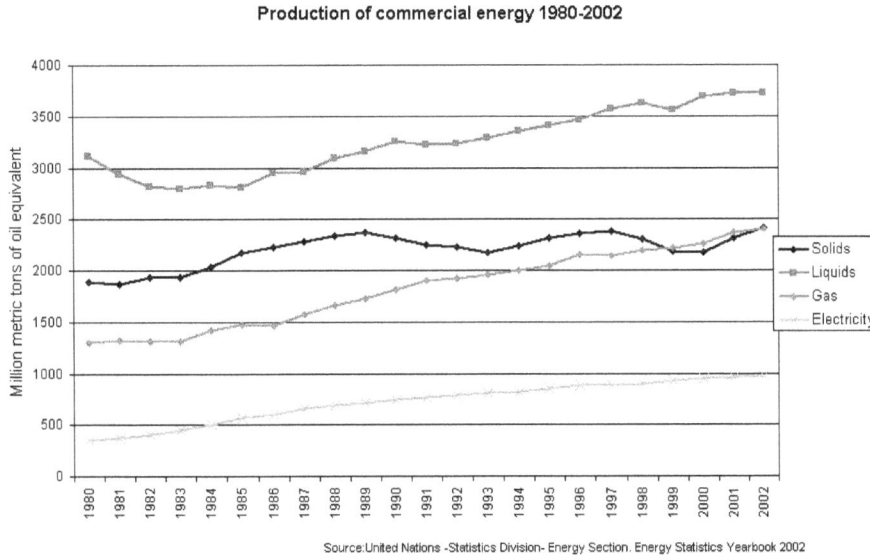

Source:United Nations -Statistics Division- Energy Section. Energy Statistics Yearbook 2002

Fig. 3.7

Source: United Nations Statistics Division. (2003). Energy Statistics Yearbook 2003-Tables (Graph: Production of commercial energy 1980-2002) [WWWpage]. URL http://unstats.un.org/unsd/energy/yearbook/EYB_pdf.htm

Evolution of Aluminium Recycling in Western Europe* 1980-2004

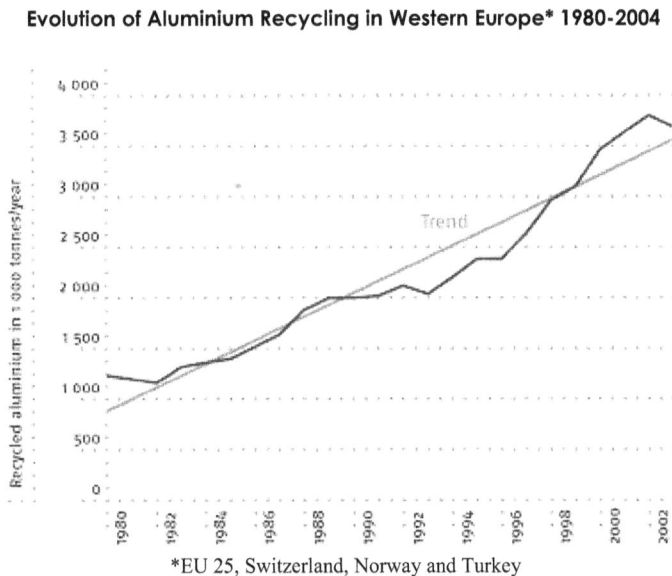

*EU 25, Switzerland, Norway and Turkey

Fig. 3.8

Source: European Aluminium Association. (2005). Recycled Aluminium (Graph: Evolution of Aluminium Recycling in Western Europe* 1980-2004) [WWWpage]. URL http://www.eaa.net/home.jsp?content=/material/recycled.htm

Bar graphs

Bar graphs show the relative size or volume of discrete variables as opposed to the continuity and trend depicted by line graphs. They use vertical or horizontal bars to show similarities and differences not only between large and small numbers but also between similar numbers. They are useful for illustrating magnitude effectively and for contrasting variables. Nevertheless, they do not show well the absolute values of the variables studied, although this can be solved by marking these values on the graph. Vertical bar graphs also function well for illustrating discrete values over time.

	R+D jobs/100 Mpeople	R+D investment (k$)/R+D jobs	R+D staff with degree/R+D staff (%)
Germany	59	76,8	48
Spain	19	63,5	58
France	55	84,1	46
Italy	25	89,5	52
United Kingdom	48	77,8	50

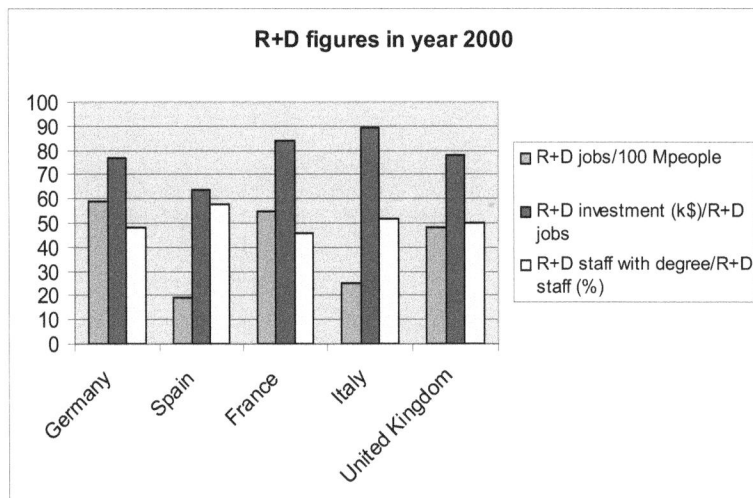

Fig. 3.9

Pie charts

Pie charts work best for illustrating portions which total 100 percent. They are also useful to compare fairly dissimilar magnitudes, in order to show gross differences, and to compare sections to one another and to the whole. Nevertheless, a pie chart shouldn't include too many segments, as the relationships among more than five or six divisions are difficult to see and may confuse readers. Also note that the difference between close percentages may not be easily distinguished, in which case it is particularly advisable to provide the percentages for each of the wedges. Likewise, label absolute values if you want to highlight them.

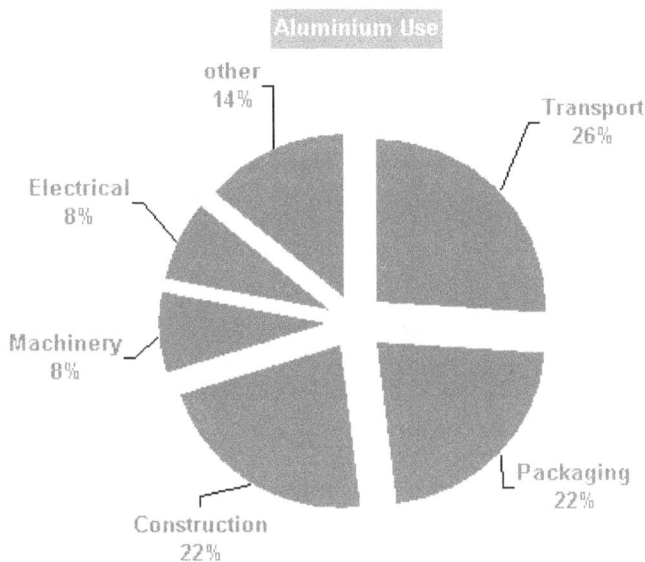

Aluminium Use

other
14%

Transport
26%

Electrical
8%

Machinery
8%

Packaging
22%

Construction
22%

Fig. 3.10

Source: BaseMetals.com.   (n.d.).   Aluminium.   (Graph: Aluminium   use)   [WWWpage].   URL http://www.basemetals.com/html/alinfo.htm

Line drawings

Line drawings include drawings of models and objects of science and engineering which focus essentially on physical appearance. Their main purpose is to show the different parts of a mechanism or to emphasize a particular shape or part (see figure 3.11), function or performance (see figure 3.12). They are preferred to photographs because they allow you to give more detail and to highlight components that couldn't otherwise be easily distinguished due to their size or complexity. Cross sections may also be provided to show both the interior and exterior or the front and side views of the same object at the same time. Obviously, the drawings should be clear enough and maintain correct proportions so that the reader does not miss any important detail. It is also helpful to use call-outs or labels to name the different components so that readers can easily identify them. As an example, consider figure 3.11 below.

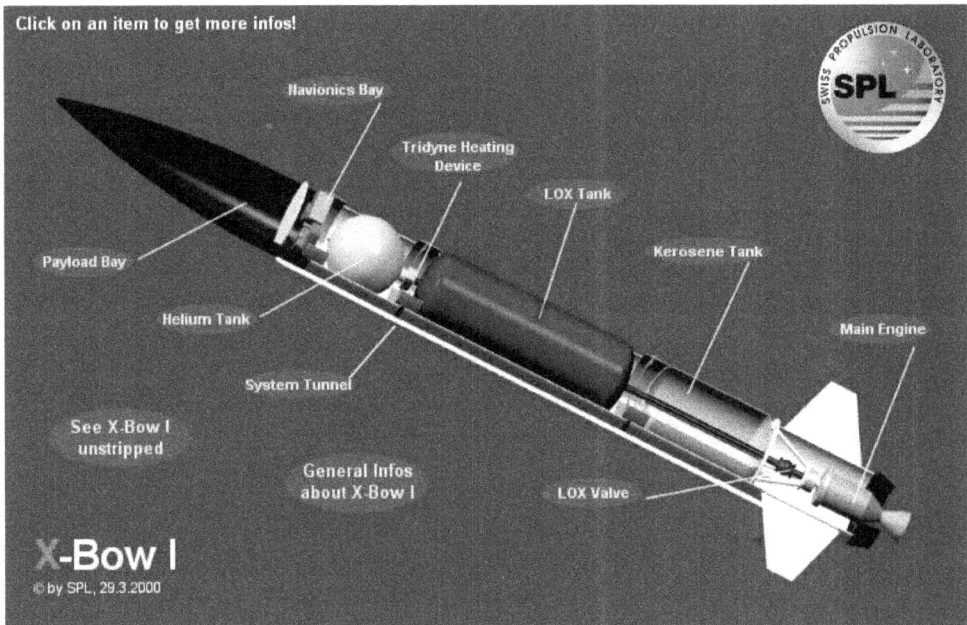

Fig. 3.11

Source: Swiss Propulsion Laboratory. (2000). X-Bow I.(Line drawing: X-Bow I) [WWWpage]. URL http://www.spl.ch/

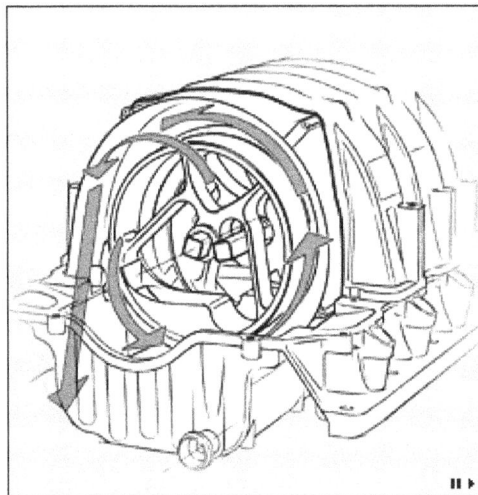

Fig. 3.12

Source: BMW. (2004). The comprehensive BMW Technology Guide. (Photograph: Fully-variable intake system (DIVA)) [WWWpage]. URL http://www.bmw.com/com/en/index_narrowband.html

Photographs

Photographs are useful to illustrate the text effectively and truthfully as they are a reproduction of what the text describes. They work well when you are trying to emphasize the external appearance of the object described as opposed to its internal composition or cross section. They show the object as a whole and emphasize all the parts equally. In fact, the sometimes excessive and unwanted amount of detail and an occasionally deficient reproduction constitute the two main drawbacks of photographs. Therefore, if you intend to focus on a particular aspect or component either label the corresponding parts or use a drawing instead.

Fig. 3.13

Source: BMW. (2004). The comprehensive BMW Technology Guide. (Photograph: The turbocharger) [WWWpage]. URL http://www.bmw.com/com/en/index_narrowband.html

Others

Apart from the graphics described above, there are some others that may also help you present information more clearly such as icons, schematics, flowcharts and pictures or cartoons. Icons offer you a solution when you need to provide a representation of a danger, prohibition, direction, etc. They should be simple, realistic and easily recognizable in order to communicate an idea rapidly and effectively (see figure 3.14). Schematics are commonly found in electronics and engineering fields to present abstract information precisely. They use highly technical symbols and abbreviations that can only be understood by experts (as an example see figure 3.15). Flowcharts are used to show a chronological sequence of activities where the different steps and options are marked by geometric figures. Remember that an oval represents a start or a stop, a rectangle describes a step and a diamond implies making a decision (see figure 3.16). Finally, humorous pictures and cartoons (see figure 3.17) help you enliven your text and break its uniformity. At the same time, they arouse readers' interest and help them remember the topic better. However, they are not commonly found in all types of documents; for example, textbooks, popular science articles and some manuals may include some but they are rare in research articles, letters or formal reports. Even if you decide to make use of them, don't use them extensively as they may distract readers and hinder comprehension.

| Corrosive substance class 7 | general prohibition | use safety harness | radioactive material |

Fig 3.14 Icons

Source: CPTec words and pictures. (2006). Safety symbols and pictograms: technical usage. [WWW page]. URL http://www.cptec.org/symbols_technische_hinweise.html

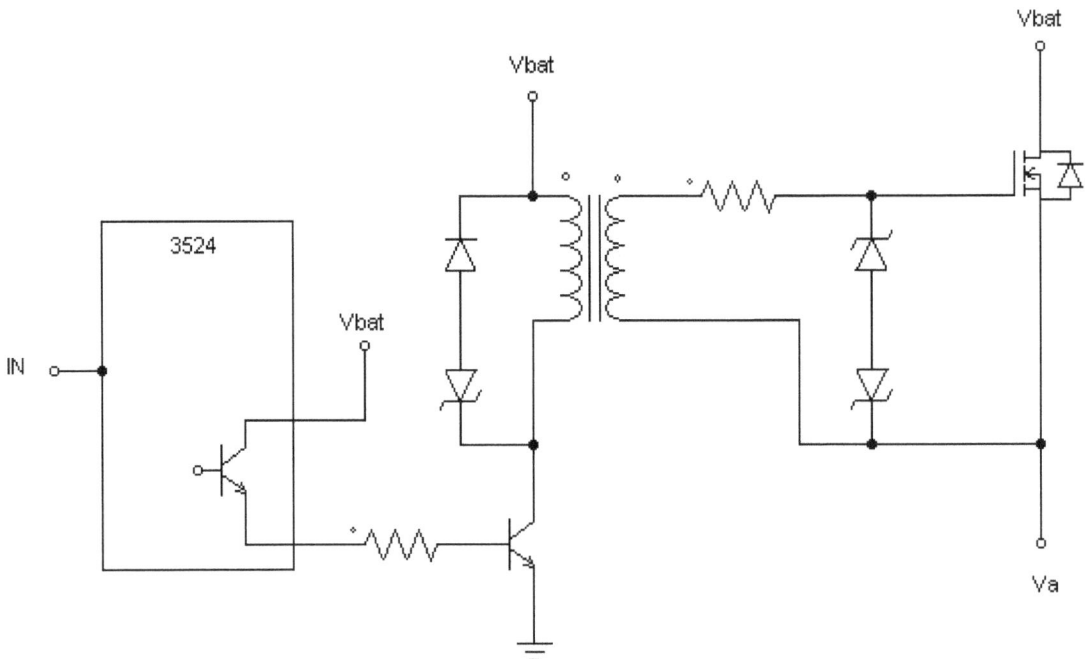

Fig. 3.15 Circuit schematic

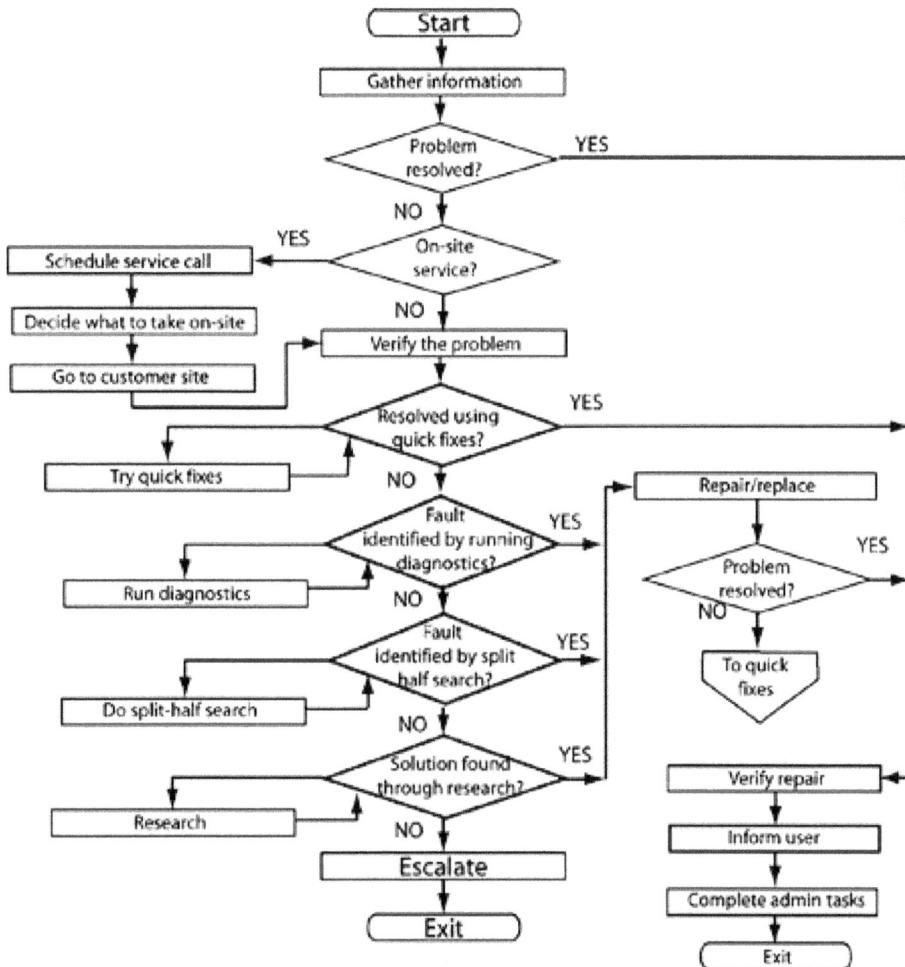

Fig. 3.16 Flowchart (© Apple Computer, Inc.)

Nowhere is the impact of China's growth clearer than in the world's commodity and raw materials industries. No industries will lose more if that growth slows.

Fig. 3.17 Picture

Source: Muñoz, C.. (2004). Cartoon: The hungry dragon. [WWWpage]. URL http://www.economist.com/business/displayStory.cfm?story_id=2446908

Designing a visual element

As explained, visual aids are a key element in technical writing. They help you summarize and explain data, present facts and figures in an orderly way, illustrate content and sustain readers' interest. However, if you want them to be effective, you should select the most appropriate type of visual aid for each occasion and make the visual aid truly visual. Below you have some major guidelines that may help you create a successful visual element:

- *Plan a relevant visual aid.* Ensure that the selected visual aid is the best option to make the point you intend to transmit. Analyze the strengths and weaknesses of the different types of visual elements before deciding on one and then carefully choose the information it should include. Avoid superfluous or redundant information and make your graphic match your text so that a reader can easily follow your train of thought. If the visual element includes contradictory, imprecise or unrelated information, it may confuse readers and lose its original communicative purpose. Remember that a good communicative visual element should always be a helpful complement that adds constructively to the text.

- *Do not distort data.* You may distort data in two different ways; namely, by providing inaccurate information and by choosing a poor display. Always check that your sources of data are reliable and that the data provided is truthful, otherwise your arguments will be considered poor and unconvincing. On the other hand, when displaying data, graphics should reflect reality. A misleading choice of visual aid or of data arrangement may distort information and obscure your point, so be extremely careful with data display and management.

- *Design a clear visual aid.* After deciding on the most appropriate type of visual aid and on the data to include, you should provide the information that allows a visual element to be independent from the text and to be read by itself. That is, include a numeral and a title with the contents of the graph, label parts and segments when necessary, provide appropriate headings and brief but informative titles and leave enough white spaces in order to provide a clear picture. In a word, include all the necessary explanatory details and don't leave anything up to the reader.

INDIVIDUAL TASKS

3-46 Convert the following text into a table so that readers can clearly see the data comparative relationships between China and India that the text points out. When designing the table, remember to include an appropriate title identifying the main point of the table and also short and informative column headings.

India lagging in science and technology, says official

T. V. Padma
29 August 2006
Source: SciDev.Net

[NEW DELHI] A senior official in India has warned that the country is lagging behind others in both research funding and scientific productivity. Speaking at a press briefing on Friday (25 August), T. Ramasami, secretary of the Department of Science and Technology, also outlined policy plans to reverse the trends. India invested US$3.7 billion in science in 2002-03, he said. The figure is dwarfed by the US$15.5 billion spent by China, US$124 billion by Japan and US$277 billion by the United States. Ramasami said India's scientific competitiveness, as measured by the number of publications in research journals listed by the Science Citation Index, is not proportional to its inherent strength in science. China, another highly populated Asian nation that has achieved considerable economic growth recently, is racing ahead (<u>see table</u>). In 2002-2003, its researchers published 50,000 papers in journals listed by the index, compared to 19,500 by Indian scientists. This is also due to the number of workers in research and development in India, only 115,500 in comparison with the 850,000 working in China. This also affects the number of doctorates produced per year in India and China, 4,500 and 40,000 respectively.

"We cannot afford to be seen to be lagging behind," said science minister Kapil Sibal. Ramasami said India has planned a series of interventions that could improve India's global Science Citation Index ranking from 14th to 7th (in terms of percentage share of global publications India falls to 1.9% behind China with 5%). These include increasing investment in science four-fold, from 250 trillion rupees (US$5.4 billion) today to US$21.5 billion in five years' time. India also plans to attract talented school students to careers in science before they lose interest in the subject. It aims to provide the top 500 students with guaranteed funding from the age of 17 through to 32 so they can pursue their education and early careers in research. In addition, India will organise scientific summer camps attended by Nobel laureates and famous Indian scientists for the top one per cent of school leavers.

Source: SciDev Net. (2006). India lagging in science and technology, says official [WWWpage]. URL http://www.scidev.net/News/index.cfm?fuseaction=readNews&itemid=3072&language=1

3-47 With the data below, design a table, a bar graph, and a line graph. Then, describe briefly the situation for which each graph would be most appropriate. Finally decide which would be the most suitable options in this particular case.

Sales of 4 different brands of laptops in a particular area over three consecutive months:

Laptop A	250,000 items (April)	200,000 items (May)	150,000 items (June)
Laptop B	200,000 items (April)	180,000 items (May)	190,000 items (June)
Laptop C	100,000 items (April)	150,000 items (May)	200,000 items (June)
Laptop D	50,000 items (April)	70,000 items (May)	75,000 items (June)

3-48 Create a chart for any information of your interest. Based on the information you want to show decide on the type of chart that best displays your data relationships. For example, if you want to show proportions or percentages try a pie chart. If you want to compare values, try a bar chart or if you want to show changes through time use a line chart.

COLLABORATIVE TASKS

3-49 Look for examples of good and bad visual aids in textbooks, journals and scientific magazines. In small groups, discuss their strong and weak points justifying your answers.

Then, each group should choose a couple of visual elements that reveal more deficiencies and decide how they could be improved. Finally, present the results of your revisions to the rest of the class.

3-50 In groups, think of an instructional process you know well—for example, a construction or design process, troubleshooting instructions, checking and overhauling procedures—and then design a flowchart describing it. Remember to use the different flowchart symbols appropriately.

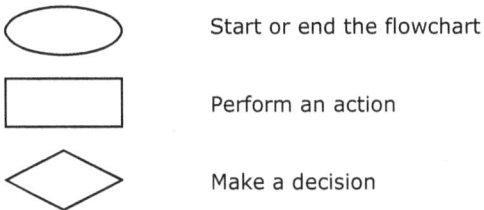

Start or end the flowchart

Perform an action

Make a decision

3-51 In groups, create new icons for the different university rooms and spaces—library, classrooms, study rooms, cafeteria, dean's office. Don't use any icon seen before; then share your options with the rest of the groups and decide on the best option for each particular case.

3-52 In groups, create new icons for the following functions and situations: harmful to the environment, hand crushing hazard, lifting hazard (heavy object), hot surface, emergency exit. Don't use any icon seen before; as in the previous exercise, share your options with the rest of the groups and decide on the best option for each particular case.

PROJECT

Now that we have completed the writing stage, you should carry on with your writing project. At this stage, you will have to draft your text in order to structure and format it in a coherent way. Remember that at this stage you shouldn't worry about grammatical or stylistic errors. You can correct them during the post-writing stage.

Bottom-up approach

STEP 3

Once you have defined your audience, determined your purpose, chosen the most appropriate style and tone, gathered information and drawn an outline, you should *begin with the writing itself*. Now it is time to expand the ideas sketched in the outline into full sentences and paragraphs. You may begin by developing a rough draft taking into account all the decisions taken during the pre-writing stage. During the writing stage, mainly concentrate on structuring and formatting your text and making it coherent. Thus, you should:

- *Structure your paragraphs.* Open your paragraphs with a suitable topic sentence from which the supporting sentences will stem. Also consider closing some paragraphs with a concluding sentence.

- *Use patterns of paragraph development.* Remember to organize supporting sentences consistently using some of the common patterns of paragraph development used in technical and scientific English.

- *Provide your paragraphs with intra-paragraph coherence.* Connect sentences within each paragraph so that the meaning or relationship between them is clear and easy to understand.

- *Structure your essay.* Organize information within your text in three main parts or sections, namely, introduction, body and conclusion. Remember each part has a specific function and a distinctive internal structure.

- *Use an essay pattern.* As with paragraph patterns, organize subtopics (i.e. body paragraphs) using a consistent pattern which may arise naturally from the thesis statement.

- *Provide your text with inter-paragraph coherence.* Use transitions between paragraphs to give unity to the text.

- *Incorporate visual aids.* If you want to present information effectively you will need to include visual aids, which will allow you to present a lot of information clearly and concisely and will constitute a visual appeal to the reader.

Top-down approach

STEP 3

In this approach, you have already analyzed whether your document accommodates to the intended audience and purpose, revised the tone and style you have adopted, validated the information and ideas in your text to see if you've written with authority on the topic and identified any omissions or imbalance of information. Now it is time to *check the structure, format and coherence of your text* and to rewrite those parts which may present deficiencies. Therefore, during this writing stage, you should revise:

- *The structure of your paragraphs.* Check that your paragraphs open with a suitable topic sentence from which the supporting sentences should stem. Also check whether some paragraphs need a concluding sentence.

- *The patterns of paragraph development used.* Remember that supporting sentences should be organized consistently using some of the common patterns of paragraph development used in technical and scientific English. If you spot any inconsistencies, use an appropriate pattern.

- *The effectiveness of intra-paragraph coherence.* Revise whether sentences within paragraphs have been properly linked so that the meaning or relationship between them is clear and easy to understand.

- *The structure of your essay.* Check whether information within your text is organized in three main parts or sections (introduction, body and conclusion) and revise the specific function and internal structure of each part.

- *The essay pattern used.* As with paragraph patterns, check whether body paragraphs are organized using a consistent pattern which should arise naturally from the thesis statement. Again, if you spot any inconsistencies, apply an appropriate pattern.

- *The effectiveness of inter-paragraph coherence.* Revise whether transitions between paragraphs give unity to the text and correct any weakness.

- *The visual aids included.* Check whether you have incorporated enough visual aids to present information effectively and appealingly to the reader. Also revise whether the visual aids included are clear and concise enough.

CHAPTER 4

Post-writing stage

Reflecting on...

Once you have written a document, what do you do next?

What do you think is more important: revising for content and organization, for grammatical accuracy or for stylistic aspects? Why?

Have you improved any aspect of your writing after your own reviews? Have you been able to identify any weakness?

Do you think it is useful to let a colleague review your document? How would you appreciate the feedback you receive?

4.1 Introduction

With the post-writing stage you have reached the final step in the process of writing. This stage is possibly the most important and also the most time-consuming as it entails going through different substages throughout the document; we could view it as a process within a process constituted by four substages or phases. However, the amount of time and effort you should spend in post-writing will be subject to the type and importance of the document. Writing a formal report that will be closely read by different readers with different interests and from which important decisions will be made is not the same as writing a memo or a letter to a peer in which informal information is transmitted. The first type will obviously need a multiple and careful revision whereas the second type will simply need a quick check over. Experienced writers know that writing effective documents is not an easy task and that they cannot obtain a good final version in a first draft. Precisely, they need to write several drafts and revise them before reaching a final version that fully satisfies them.

In the pre-writing stage you have identified the audience and purpose, decided the tone and style and outlined the content. In the writing stage you have written down a complete draft. Now, in the post-writing stage, it is time to polish the document, to properly *finish* it. To do so, you will have to add, suppress, modify, simplify, enhance, move the information around or make any other changes so as to ensure that you say what you really wanted to say and in the correct form and style. A poor or a lack of revision will not only cause misunderstanding but also give the impression that the document has been written carelessly, which would undermine credibility.

Post-writing, as previously indicated, can be divided into four main substages that require separate study due to their complexity and idiosyncrasy. These substages are:

- Revising content and organization
- Checking for grammatical accuracy
- Editing for style
- Proofreading and peer review

4.2 Revising content and organisation

Revising means looking over again and, in this context, it will refer to looking over content and organization. Given the existing interrelationship between these two aspects, we have included them in the same substage. The best revision is done in two separate readings, one for the content and one for the organization; this way you will be able to concentrate on each aspect better and fewer details may escape your attention. However, depending on the length of the document, you can revise at the content level and consider the organizational level at the same time, for revising one aspect may imply changes in the other.

Revising content

When revising for *content* you should check, taking as a reference the planned outline, that central information has been properly included, that superfluous or redundant information has been left aside and that the amount of details and supporting information are appropriate. Writers that are too involved in what they are writing about run the risk of going too much into detail, thus preventing important facts from standing out. Related to this issue is the idea of *proportionality*, which is based on the principle that more important facts require longer explanations.

The idea of *completeness* is also important; that is, that the introduction, the body and the conclusion include the corresponding information and that the same information is not repeated in two different sections. Besides, you should ascertain that the supporting information stems from the main ideas or topic sentences and also that all supporting sentences in a paragraph are related to the topic. Whenever you feel that a sentence has no relation either to the main idea or to any previous sentence, you should prune or reformulate it. Likewise, you should check that all paragraphs keep *thematic unity*; namely, that there is no paragraph that goes off topic. As for the conclusion, check that it follows logically from the information presented in the body. Finally, you should remember that technical writing emphasizes a logical way of thinking and building arguments so abrupt leaps of logic will not be very well-accepted by technical readers.

Closely related to the content, the *title* should also be revised. You should ensure that it is both appealing to the reader and informative in order to be effective. Although less frequently than in literary writing, where imagination has a more prominent role, important modifications can easily creep into the main text during the process of writing, having a direct effect on the initial version of the title. Alternatively, and to avoid rewriting it, you can write the title once the document is completed. Creating the title is an interesting part of the whole process and consists basically in combining key words in a strategic, significant and catchy way. Avoid writing titles that use metaphorical and cryptic expressions as in newspapers, poetry and other forms of non-technical writing. In technical writing you cannot play so much with words and expect the reader to spend time figuring out the meaning of the title. Also, remember that the title cannot be a full sentence, but a combination of key words, which sometimes consists of a two-part expression in which the first part indicates the content and the second the function or format. Some examples are:

The ultimate distributed workforce
Handling of volatile liquid fuels

Voice coding: An overview.
Tidal energy. A case study.

A reader-oriented revision will also revise the audience the document is addressed to. As reviewers, we have to check whether the content of the document satisfies the intended primary audience and find out whether a secondary or tertiary audience is also possible. If this is the case, especially the level of detail may have to be modified to satisfy these potential audiences.

The following checklist will help you revise content:

CONTENT CHECKLIST

1. Does the title clearly announce the content?
2. Is the content adjusted to the intended audience/s?
3. Do supporting or secondary ideas stem from a main one?
4. Is central information included and given prominence?
5. Is the amount of detail adequate?
6. Are more important ideas proportionally longer than less important ones?
7. Is the same information repeated in different sections?
8. Is the use of topic sentences adequate?
9. Are all paragraphs related to the main topic?
10. Does the conclusion follow logically from the information in the body?

Revising organization

The first point to check when revising the *document organization* is whether the document displays a logical organization of information. By logical organization we mean that the information must be organized in a clear and consistent way so, for example, a check should be made as to whether patterns of information organization have been used adequately. This will be essentially achieved by checking that the document, besides being coherent, keeps a logical organization of its different parts or sections. In technical documents like reports, sections and subsections cannot be arranged at random; there must be some kind of underlying logic or criterion by means of which they are organized. It must be noted that in technical writing this logical organization is already pre-established by the specific format of the document. So we can say that the logical order of most technical documents is expressed by an elaborate structure of sections and subsections. This particular manner of partitioning information is very much in accordance with technical readers' selective way of reading. As opposed to non-technical writing, which is assumed to be read from beginning to end, technical writing is aimed at readers that often read in search of specific pieces of information that interest them while skipping the rest.

Another point to check is *paragraph size* and *distribution*. Avoid writing one-sentence, or two-sentence paragraphs. Only certain documents may include introductory paragraphs, summary and transition paragraphs consisting of one or two sentences; main paragraphs should display a fair degree of development. So if you have written too many short paragraphs, consider combining them into longer ones as well as adding examples, illustrations and details to avoid underdevelopment.

You should also check that organizational divisions have been clearly marked. There is a wide range of marking techniques, which go from chunking and ordering (like headings, white spaces, section dividers, size and position of the heading and indentation) to highlighting (such as bullets, numbering, bold face, underlining and colour). Therefore, do not stick to one or two types; on the contrary, use as many as you need to end up with a clear and attractive document. Remember that different markers are not only used to make the document more appealing but, most importantly, they also serve to mark organizational aspects such as the hierarchical relationship of information. Indentation, for instance, is one way of indicating sublevels or subcategories. Markers will be of a great help for your readers. Consequently, check if you have used varied and visible markers. Nowadays good processors offer you a great variety of possibilities to help you create a visually appealing document so that it becomes more readable.

Finally, check that the introduction and conclusion appear as complete and distinctive sections. An option for the conclusion is to use transitional markers such as *in conclusion, in brief, to sum up.* With regard to the rest of the document, check that appropriate titles and subtitles precede the corresponding sections and subsections.

The following checklist will help you revise the organization of your draft:

ORGANIZATION CHECKLIST

1. Are the different parts of the document clearly and logically organized?
2. Have organizational patterns been consistently used?
3. Are marking techniques correctly used?
4. Are titles and subtitles appropriate?
5. Is the hierarchical relationship of information clearly expressed?
6. Do paragraphs have sufficient connecting expressions to make them coherent?
7. Have transitional signals been adequately included to help in paragraph transition?
8. Are the introduction and conclusion identified as complete and distinctive sections?

4.3 Checking for grammatical accuracy

After revising content and organization, it is time to check that your writing is grammatically correct. A very important step within the post-writing stage, then, is to reread the whole document concentrating on grammatical accuracy. In order to help you at this substage, we are proposing a list of grammatical points which should go into consideration in your revision of grammatical accuracy. The following list is not intended to be comprehensive; rather, it is a list of grammatical aspects that students of technical writing usually have trouble with. However, if you feel that your command of grammar is low and that very basic mistakes often creep into your writing, you should go to the handbook in the last chapter and invest some time in making up your grammatical deficiencies.

In this section we are going to focus on different grammatical points, first giving some explanation and then providing you with exercises to gain more practice. We hope this practice helps you:

- refresh your linguistic knowledge
- consolidate or learn some specific grammatical rules
- identify your linguistic needs and weaknesses
- use the information obtained to construct your own learning path

The grammatical points that will be dealt with to help you spot mistakes during your revision are the following:

- Subject-verb agreement
- Pronoun agreement
- Pronoun antecedent
- Subject repetition
- Subject omission
- Dangling modifiers
- Reduced relative clauses
- Run-on sentences
- Sentence fragments
- Nonparallel structures (revision)

Subject-verb agreement

Subjects and verbs must agree in number. This is such a basic grammatical rule that even a beginner knows that if the subject is plural, the verb is also plural:

> e.g.   Engineers are skilled workers

However, there are some cases where the problem lies in the words that act as subject or in identifying the subject. Even native speakers of English can make this silly mistake:

> e.g.   *Each of the fifteen engineers in the technical department *need* English.
> (The subject is *each*, a singular noun, so the verb must agree with the subject and also come in singular)

You should pay special attention to the following particular points which may pose certain difficulties:

Two or more (singular) subjects joined by *and* take a plural verb:

> e.g.   Bioenginering and environmental engineering *are* required courses in the current general engineering curriculum.

But when there are expressions like *together with, along with, as well as, rather than* between commas, the verb agrees with the number of the first noun:

> e.g.   Bioenginering, as well as environmental engineering, *is* a required course in the current general engineering curriculum.

In coordinations using *either...or* and *neither...nor*, the pronoun agrees with the nearest subject:

> e.g.   Either the *technician* or the *supervisor is* likely to disapprove of the procedure.
> Either the *technicians* or the *supervisors are* likely to disapprove of the procedure.
> Either the technician or the *supervisors are* likely to disapprove of the procedure.
> Either the technicians or the *supervisor is* likely to disapprove of the procedure.

When the number of the subject and the complement differ, the verb should always agree with the subject.

> e.g.   Repeated inaccuracies were the reason for the experiment failure.
> The reason for the experiment failure was repeated inaccuracies.

Each, every, neither, either, everybody, somebody, anybody, nobody, anyone, are singular subjects:

> e.g.   Every of the three hundred two-pinned-plugs *was* defective and *was* returned to the manufacturer.

None applies to three or more entities but it can be treated as a singular or a plural noun:

> e.g.   You should check that none of the components *are/is* defective.

Collective nouns can be either singular or plural[9]: *team, jury, class, committee, orchestra, board, family, group, club, union, staff, company, society, faculty, government, college, panel, association, university, crew, corporation, commission, audience, department, majority, generation, majority, etc.*

> e.g.   The research team *consists/consist* of nine mechanical engineers, two economists and one mathematician.

---

[9] In American English collective nouns are usually treated as singular.

The difference lies in the emphasis or point of view: the singular stresses the non-personal collectivity of the group whereas the plural stresses the individuality of the different members of the group:

> e.g.   The commission *haven't* been able to reach a consensus yet.
> (Attention is given to the individual opinions of members of the group)
> The committee *has met* and *has rejected* the proposal.
> (The committee acts as a single collectivity)

Non-count nouns denoting an undifferentiated mass or continuum are always singular. Examples are: *water, warmth, dampness, news, advice, information, research, work, evidence, business, equipment*. With these nouns the expression of quantity may be achieved by means of partitive nouns like *pieces of*, *items of*, *bits of*:

> e.g.   an item of news   several items of news
> a piece of advice /information /research   some pieces of advice/...
> a bit of chalk   some bits of chalk

There are some nouns mostly derived from Latin or Greek that take special plural forms instead of the usual *–(e)s* ending.

> e.g.   The *curricula have* to adapt to our ever-changing world and so environmental courses are now offered.

Could you complete the table below?

Singular	Plural	Singular	Plural
analysis	analyses	datum	data[10]
hypothesis	hypotheses	curriculum	
basis	bases	medium	
ellipsis	ellipses	memorandum	
crisis		focus	focuses/foci
thesis		nucleus	
axis		formula	formulas/formulae
phenomenon	phenomena	antenna	
criterion		matrix	

Nouns ending in *–ics* denoting subjects or sciences are usually treated as singular:
> e.g.   Genetics /economics/ electronics/ mechanics/ politics/ physics /acoustics... *is* an all-important topic.

---

[10] Although strictly speaking *data* is a plural noun, nowadays it is frequently used in singular especially in American English.

Also note that there are some words like *means* and *series* which, in spite of ending with a final –*s,* can be either singular or plural.

e.g.   A means/series *is* ───────▶ two means/series *are.*

INDIVIDUAL TASK

4-1 Now read the following sentences and decide whether the infinitive in brackets should take the singular or the plural form.

1. The printer and the plotter we have bought (be) the best choice for us.
2. The new equipment is necessary to obtain the items of information that (be) going to buttress our research.
3. Gathering information without resorting to plagiarism (be) not easy in the age of the Internet.
4. The CD, together with the DVDs, (have) been stolen from the filing cabinet.
5. Everybody (know) that the economics of the timber trade (be) ruling that country.
6. Versatility in engineers (have) proved to be highly appreciated among employers.
7. The analysis he made (be) rejected because the criteria he had chosen (be) neither specified nor homogeneous.
8. Provided the data (have) been corrected and the hypothesis of the work (have) been adapted to our goals, the government may sponsor the research.
9. The decision of using recyclable materials and of modernizing the coffee facilities in the office (have) been welcomed by the staff.
10. Frequent complaints (be) the battle that the quality department (have to) win.
11. What (need) further analysis (be) the pitfalls of being computer-illiterate.
12. An appalling series of coincidences (have) apparently contributed to the success of the gadgets which (sell) so well.
13. The board (have not) been able to reach an agreement on how to fund the telecommunications programme, but either private companies or the government (be) likely to be the main sponsor.
14. A sum of EUR 1000 (be) a lot of money.
15. The heating of the water is carried out without any direct contact with other liquids, which (prevent) clean water from being polluted.

Pronoun agreement

This mistake is closely related to the previous one. Just as verbs agree with their subjects so do pronouns with their antecedents, not only in number but also in gender. In terms of grammatical complexity, it is therefore very simple. Pronouns always refer to or substitute nouns, other pronouns, or phrases, and the word they refer to is known as 'antecedent'. When revising your document you have to check that every pronoun agrees with its antecedent, as in the cases below:

> e.g.   In the pre-writing stage, the audience *is/are* to be analyzed as well as *its/ their* level of technicality.
> (When the subject is a collective noun, you must be careful to be consistent with the concordance between the number chosen for the subject and the pronoun.)

> e.g.   Everybody knows what *his /her* duties are.
> (The pronoun *his/her* refers to the antecedent *everyone*, which we said is a singular noun.)

Writers willing to avoid sexist language may face the problem of which pronoun (*she/he* or *her/his*) they should use. One solution is to use both the feminine and the masculine pronouns separated with a stroke; another solution is to use the plural form.

> e.g.   *Each* applicant will have to submit *his/her* CV by the end of the week.
> e.g.   *Applicants* will have to submit *their* CVs by the end of the week.

INDIVIDUAL TASK

4-2 Fill in the blanks with the correct pronoun, making sure each pronoun agrees in number and gender with its antecedent. Then, underline the correct verb tense, making it agree with its subject.

1. Neither of the two computers (has/have) been moved to............right place.
2. After two weeks of hard work, the team (has/have) announced that.............will soon publish the results of.............research.
3. The two technicians redesign the metal structure of the partition walls separately and then each (puts/put) forward.............proposal for improvement.
4. The university should reconsider...........performance evaluation system.
5. The Board receives complaints, analyzes them, and reaches............verdicts. Once reached, (this is/these are) publicly announced.
6. The hypotheses and............corollary assumptions (has/have) affected the conclusions drawn.

7. There are many advantages and disadvantages of alternative sources of energy people should know and the media (needs/need) to show some concern about............. .

8. The committee will issue............yearly publication including all of the scholarship awards given.

9. One should try to introduce resourcefulness and eliminate monotony from..........work.

Clear pronoun antecedent

Using a pronoun whose antecedent is unclear is a pervasive mistake in students' writing. Pronouns like *it, they, which, this,* or *these* must refer clearly to their antecedent; otherwise, the interpretation of the sentence may be ambiguous. A pronoun whose antecedent is vague may even give rise to misinterpretations. Several factors can cause a pronoun to be ambiguous: lack of proximity, unstated antecedent, and abstract or indefinite antecedent. In these cases, it may be useful to repeat the antecedent or rephrase the whole sentence. Below are several examples with an unclear pronoun antecedent:

(*Ambiguous*)   In the department, *they* said that people willing to be relocated only needed to fill in a form.
(*they* has no stated antecedent)

(*Revised*)   The department informed us that...
or: In the department, we were informed that ...

(*Ambiguous*)   Peter told Mr Jackson that *his* letter had been returned.

(Although the principle of proximity says that a pronoun antecedent is the nearest noun, *Mr Jackson*, in a sentence like this *his* is an ambiguous pronoun as it may also refer to the first noun *Peter*. Some possible ways to avoid this ambiguity are shown below.)

(*Revised*)   *Peter* told Mr Jackson that *Peter's* letter had been returned.
Peter told his chief *Mr Jackson* that *his chief's* letter had been returned.

(*Ambiguous*)   The human resources manager unveiled in a press release the reasons why 300 workers had been made redundant. *This* was criticized.

(*Revised*)   The human resources manager unveiled in a press release the reasons why 300 workers had been made redundant. *This public revelation* was criticized.

(*Revised*)   The human resources manager unveiled in a press release the reasons why 300 workers had been made redundant. *The reasons for this dismissal* were criticized.

INDIVIDUAL TASK

4-3 Read the sentences below, identify any ambiguous or wrong pronoun and circle its antecedent. Making as many changes as necessary, you should then provide a better alternative.

1. Even though Max is a very good product designer, he's never taken a lesson in it.
2. Absenteeism was such a worrying fact that they decided to penalize it.
3. The chief engineer Mr Higgins wished his computer had saved the information in his memory before the lights went off.
4. In the research article, it claimed that the latest fatigue resistant composite had in fact negative long-term consequences.
5. Clara's colleague told her she shouldn't have refused the promotion.
6. A leg of a robot was thrust into the air, hitting the microscope and breaking the test tubes next to it. Strange though it may seem, it was only slightly damaged.
7. The two civil engineers, acting on behalf of their company, made a formal complaint about the irregularities in the tunnel construction, which irritated the mayor.
8. The two tunnel construction methods have been widely used around the world. The first is said to be cheaper and only slightly less safe than the second. Therefore, it is not clear why the builder cannot choose it.
9. The first ever-created multi-disciplinary research group of doctors, biologists and engineers working on genetics is hovering on the brink of a new breakthrough. This is expected to benefit humankind.

Subject repetition

This is a typical mistake among Spanish writers when they write certain impersonal statements in English. Spanish writers sometimes make this mistake when they write passive sentences and want to translate impersonal expressions like *se coloca un aislante*. You have to bear in mind that a passive construction, like an active one, always needs first a subject and then a verb. Consider:

e.g.	*It is placed an insulating material.
(Revised)	I place an insulation material. (active)
(Revised)	An insulation material is placed. (passive)

This standard passive sentence is not to be mistaken for another impersonal form, always prefaced by *it*. Remember that these "introductory *it*" statements, which also allow you to be impersonal, typically have either one of these two patterns:

<u>Pattern A</u>: It +be + adjective + to + infinitive

e.g.   It is unlikely/difficult/advantageous/… to find contradictory results.

<u>Pattern B</u>: It + past participle (passive verb) + that + clause

e.g.   It has been demonstrated that excessive exposure to microwaves from portable phones can be carcinogenic.

INDIVIDUAL TASK

4-4 Correct the sentences by making as many changes as necessary. Note that there may be more than one mistake in each sentence.

1. It is believed that it can be triggered the alarm simply because of thunder.
2. It is described the system in the first part of the chapter and in the remainder statistical analyses are performed.
3. It is feasible to build the new laboratory this year, as it is financed the construction by the autonomous government.
4. Recently it has been performed studies whose aim was to obtain analytic expressions that allow the study of the rectifier behaviour in an easy way.
5. As soon as it is installed the chip, the unit is expected to operate properly again.
6. In this central heating system in particular, it is also very important the pump because it makes water flow through the pipes and radiators.
7. It was estimated twenty-one vacancies in the department.
8. The university researchers have been given the permission to carry out the research project which it will be sponsored by the local administration.
9. They were analyzed the samples and the result that it was obtained from the statistical analysis was checked.
10. Finding alternative sources of energy and familiarizing the population with them it is not an easy job.

Subject omission

This type of mistake is closely related to the previous one. Because in English sentences must always have a subject, it is incorrect to begin a sentence with the verb, leaving out the subject. If the subject is impersonal, remember you can use the passive or either of the two patterns we saw in the Subject repetition sub-section, i.e. an impersonal statement with an "introductory *it*".

e.g.   *Is impossible to deactivate the alarm without damaging the system.

(Revised)	*It* is impossible to deactivate the alarm without damaging the system.
(Revised)	*The alarm* cannot be deactivated without damaging the system.

INDIVIDUAL TASK

4-5 Read these sentences and make the necessary changes to make them grammatically correct. Note that there may be several ways to correct them:

1. Exist two alternative factory layout designs but we are waiting for the environmental engineers to issue their feasibility report.
2. Has been proved that the chemical factory is disposing of hazardous waste and throwing it into the river.
3. The seatbelt is part of the safety equipment in a car; if well used, is very important because can prevent you from being seriously injured in a car accident.
4. Since the pipes are usually made of metal and since water can produce corrosion on them, is advisable to check the installation once a year.
5. It is known that exists the problem of keeping all rooms in a house at the same temperature.
6. Hold the mouse with your right hand. As you hold it is necessary for you to look at the monitor while you are moving the small device.
7. Installing a door switch alarm system at one's home is a good idea. If a thief broke into the house, would activate the alarm and is very likely that he/she was caught by the police.
8. Nowadays is arising an important problem in our planet: the continuous exploitation of natural resources has caused an increase in pollution and an alarming shortage of these resources.
9. In the process of selecting the parts of a water pumping station are two options—a diesel engine or a gasoline engine.
10. With regard to fuel cells as a type of power technology, is hoped that efficiencies of over fifty percent can be achieved in the next ten years.

Dangling modifiers

Dangling modifiers, also known as *misrelated participles* or *misplaced modifiers*, usually prove troublesome to many students and some may give rise to unintentionally humorous messages. There are two possible types of modifiers: unclear present and past participles and wrongly located phrases or nouns.

Dangling modifiers: type 1

The first type of dangling modifiers are basically present or past participles whose subject is absent. If you do not want to misuse modifiers, it is very important that you regard the following principle: the subject not present in the non-finite clause should coincide with the subject of the independent clause. Otherwise, the non-finite clause will be said to be dangling.[11]

> e.g.*Worried* about the absenteeism rate, *Mr Edmond* wrote an email to the personnel manager.
> (The past participle, *worried*, refers to Mr Edmond)

It may be helpful to check that the subject of the independent clause is appropriate for the verb in the non-finite clause; that is, both subjects should be human or non-human, otherwise confusion may arise:

> e.g.   * *Repairing* the electrical installation, *the lights* went off.
>   (The lights were not repairing!)
>
> (Revised)   When he was repairing the electrical installation, the lights went off.

Sometimes confusion may also arise when you write a sentence which is grammatically correct but may not convey the meaning you want to express:

> e.g.   In spite of disagreeing, the boss made the workers work in shifts.

This sentence is grammatically correct and it means that the boss is the one who disagreed. However if you want to say that the workers are the ones who disagree, you have to write a different sentence:

> (Revised)   In spite of *the workers' disagreement*, the boss made them work in shifts.
>   (The modifier has been replaced with a noun)
> (Revised)   In spite of *disagreeing*, the *workers* were made to work in shifts.
>   (The subject of the modifier is now 'workers', which is the intended meaning).

---

[11] There are some dangling modifiers whose acceptability is related to how easily the particular reader can perceive the implied subject, as in the following example:

> e.g. Leaving the faculty, a bright idea came to me. (Presumably the implied subject is *I* despite not occurring in the independent clause. However, the sentence may be acceptable although a better option is to rewrite it introducing *I*.)

Dangling modifiers: type 2

Dangling modifiers in the second group are usually phrases which are badly located.

e.g.   *Science students don't dare to hand in assignments to teachers *with writing problems.*
(Is it *teachers* who have *writing problems*? The phrase *with writing problems* has to be relocated and has to follow its antecedent, *students*).

(Revised) Science students *with writing problems* don't dare to hand in assignments to teachers.

INDIVIDUAL TASK

4-6 Each sentence contains a dangling modifier. Identify it and make the necessary corrections to make the sentence grammatically and semantically correct.

1. Writing a memo to his superior, a mistake was made concerning the total amount of money.
2. Towed behind the tug, I saw the oil tanker entering the harbour.
3. When writing a document, care must be taken not to misplace modifiers.
4. Materials typically used for the integrated circuit (IC), an IC is bonded to the leadframe plate using either a eutectic solder or an epoxy resin.
5. Without a mobile phone, it was impossible to communicate in that area.
6. If the air-conditioning device is to be installed, the technician should make a hole in the wall that allows cooling the room.
7. Checked by the quality manager, nobody realized that the mechanisms were not working properly.
8. The first table has to be proofread in the results section of the article.
9. Broken by some angry workers, nobody knew how to mend the video camera.
10. Ignoring the price, the latest model of digital cameras was bought.
11. After analyzing the method used, the results turned out to be unconvincing.
12. The experiment failed not having followed the instructions carefully.
13. The principal types of operation used are: sawing, drilling, routing, beveling, bonding, and forming to work laminated plastic sheets.

Reduced relative clauses

Sometimes, and particularly if you realize that you have been writing excessively long sentences, reducing relative clauses may be a good way to lighten your text without altering the meaning. Shortening a relative clause is not a difficult matter although sometimes it may pose grammatical problems. Reducing a relative clause basically consists in:

a) omitting the wh-pronoun and the auxiliary verb 'be'
b) rewriting a simple present tense into a present participle when the meaning is active
c) reducing a passive verb phrase into the participle

a) Engineers *who are working in infected areas* should be vaccinated.
   (*Reduced*) Engineers *working* in infected areas should be vaccinated.

b) Engineers *who work in infected areas* should be vaccinated.
   (*Reduced*) Engineers *working* in infected areas should be vaccinated.

c) Engineers *who have already been infected by the virus* needn't be vaccinated.
   (*Reduced*) Engineers *infected* by the virus needn't be vaccinated.

There are, however, two related issues that should be addressed. The first is that when a reduced relative clause is so simple that it merely consists of a participle, this is usually relocated and placed before the subject noun.

e.g. Engineers *who are working* should be vaccinated
     (*Reduced*) *Working* engineers should be vaccinated.

No phrase is hanging from the relative clause so we can place the present participle before the head noun, treating it as an adjective. However, this relocation is not possible if the relative clause is long and contains complements. Thus, when shortening a relative clause, you should be careful not to misplace the complement; i.e., you should not separate complements from the relative clause to which they belong.

e.g. Engineers *who are working in infected areas* should be vaccinated.
     * *Working* engineers *in infected areas* should be vaccinated.
     (The phrase *in infected areas* is dangling.)

The second point to be made concerns the fact that *not all* relative clauses can be reduced. Relative clauses containing a modal verb *can* be reduced but at the cost of losing the extra meaning of the modal verb. In brief, these relative clauses should not be shortened:

e.g. The report includes a list of experiments *which should be done this year*.
     (*Reduced*) The report includes a list of experiments *done this year*.
     (This reduction is possible grammatically speaking, but the meaning of the original sentence has been altered because the meaning of *should* has been lost.)

INDIVIDUAL TASK

4-7 Read the following sentences and join them by means of a reduced relative clause where possible. Note that there may be more than one way to combine them.

1. The drilling machinery has finally been installed. It costs EUR 12 300.
2. One of the moons revolves around the planet. The moon was photographed.
3. People who live near airports often suffer from heart problems. They are caused by noise pollution.
4. The island is being rebuilt thanks to international help. A tsunami destroyed the island.
5. Candidates who are interested in the vacancy may submit their curricula vitae. All curriculum vitae should arrive by the end of the week.
6. Pieces of equipment require intensive care. They carry a 3-year guarantee.
7. The printer is overused by the staff. The printer must be urgently replaced.
8. A new research laboratory is being built in the campus. This facility is fitted with the most advanced technological equipment.
9. The pipe had been leaking for hours. It was repaired overnight.
10. The installation was completed only a year ago. It already needs repairs.
11. The reports will be discussed tomorrow. They have been reviewed by my superior.
12. Nowadays libraries are subscribed to many journals. The journals may offer only an electronic format.

Run-on sentences

Run-on sentences have to do with punctuation. When two independent clauses appear without any punctuation, we say that we have a run-on sentence. The cumulative effect of run-on sentences can seriously hinder comprehension, so they must be corrected using any of the techniques below:

*The LaserJet 1320 printer is one of the fastest and most reliable models we have released in the market it outsells similar printers from other manufacturers.

Technique 1: use a co-ordinator or a subordinator.
(*Revised*): The LaserJet 1320 printer is one of the fastest and most reliable models we have released in the market, *so* it outsells similar printers from other manufacturers.

Technique 2: use a semi-colon or a full stop to separate the two independent clauses.
(*Revised*): The LaserJet 1320 printer is one of the fastest and most reliable models we have released in the market; it outsells similar printers from other manufacturers.

INDIVIDUAL TASK

4-8 Read the following statements and correct the run-on sentences you are able to identify.

1. Three hundred immigrant workers worked non-stop, the five-star spa hotel was built in about two years.
2. A zinc coating has been proved to reduce resistance, therefore it is no longer used in vehicles nowadays.
3. The frame of mountain bikes has not changed very much over the years, in the last few years frames have increasingly been made of lighter materials like aluminium.
4. Engineering students used to make two-dimensional drawings by hand, nowadays engineering students make three-dimensional drawings with the help of CAD programs.
5. Peter's father approves of his son's decision, however I know that underneath he would like his son to follow him into the engineering business.
6. Poor quality and lack of rigour angers my boss, he believes these attitudes have permeated the system.
7. The company is about to launch a new protective foam in compliance with the European regulation and meeting all European safety standards, it is expected to be the money-making product of the company.
8. The new legislation offers EUR 2 billion to finance solar energy projects, consequently the number of companies interested in these projects is expected to boom.
9. A newly issued report reveals in facts and figures what the average citizen has known for years, the inflation rate has been steadily increasing since the Euro first appeared.
10. You should try to proofread all your documents for run-on sentences and sentence fragments you should also proofread them for any other grammatical error.
11. Erasmus students willing to access the intranet of virtual courses should enter their user's name, which is their identity card number, enter their password, which is their date of birth plus two more digits, click on elective virtual courses and enter the intranet there they will find exercises, texts and other documentation together with the teacher's instructions.
12. You had better do the homework and send it via intranet before the deadline, afterwards you will receive an acknowledgement of receipt from the teacher.

Sentence fragments

Like run-on sentences, sentence fragments also have to do with punctuation. A sentence must have at least a subject and a verb and it must be able to stand on its own as a finished and complete idea[12]. We could then define a sentence fragment as a sentence that lacks a subject, a verb, or both; sentence fragments are usually the result of a wrong punctuation that

---

[12] For emphatic or literary purposes, in advertising or in fiction this rule may not be observed.

fragments the sentence. Below are typical examples of sentence fragments followed by the techniques used to correct them:

Problem: A dependent clause cannot stand on its own

e.g. *Tim decided to study engineering. *Because he had always loved engines and mechanisms of all kind.*
e.g. *Tim is the one who decided to study engineering. *Who had always loved engines and mechanisms of all kind.*

Technique 1: Join the two clauses with a connector

(*Revised*) Tim decided to study engineering because he had always loved engines and mechanisms of all kind.
(*Revised*) Tim had always loved engines and mechanisms of all kind, so he decided to study engineering.

Technique 2: Write two independent clauses separated by a semi-colon or a full stop, or make the relative clause a non-defining one between commas

(*Revised*) Tim is the one who decided to study engineering. He had always loved engines and mechanisms of all kind.
(*Revised*) Tim, who had always loved engines and mechanisms of all kind, is the one who decided to study engineering.

Problem: Non-finite clauses (-ing/-ed modifiers and infinitives) and phrases are not complete independent sentences and cannot stand on their own without a verb

e.g. *Tim loved engineering. *Playing with engines and mechanisms of all kind.*
e.g. *Tim had always loved engines and mechanisms of all kind. *An excellent engineer.*
e.g. * Tim decided to study engineering. *To understand how and why engines and mechanisms of all type worked.*

Technique 3: Write two independent clauses or combine the fragment with a complete sentence

(*Revised*) Tim loved engineering. He was always seen playing with engines and mechanisms of all kind. Or: Having always played with engines and mechanisms of all kind, Tim decided to study engineering.
(*Revised*) Tim was an excellent engineer. He had always loved engines and mechanisms of all kind. Or: An excellent engineer, Tim had always loved engines and mechanisms of all kind.
(*Revised*) Tim decided to study engineering. He wanted to understand how and why engines and mechanisms of all type worked. Or: Tim decided to study engineering in order to understand how and why engines and mechanisms of all type worked.

INDIVIDUAL TASK

4-9 Read the following statements and correct the sentence fragments you are able to identify.

1. High achievers are usually ambitious people. Who also like taking risks.
2. The lecturer identified three main stressors among his students; exams, lab reports and essays, and oral presentations.
3. Women engineers do not always have the same career opportunities as men engineers. Even though private and public companies do not always accept this reality.
4. Delving into the cause-effect analysis of the problem. They realized they had initially underpinned loss of ductility as a side effect while in fact it was a cause.
5. One of the greatest engineering works undertaken in Europe, this suspended bridge was built in a record time. Less than 2 years.
6. Following her lawyer's advice; Laura decided it was worth spending time and money and patented her invention last year.
7. Ms Edwards was initially reluctant to change her department organigram but she was finally convinced by Jim Leeds. A highly reputed professional.
8. These laptop computers stand out as the cheapest and most reliable in the market. Since they have been manufactured at a very low cost near Shanghai.
9. Owing to the increase in scholarships. A greater amount of students have applied for economic help this year.
10. The economist announced that Singapore, India and China could be singled out as three emerging economic forces. A widely accepted opinion.
11. If the chief computer engineer had had no fear of more viruses invading the company's net; no special measures would have been taken.
12. The architect is known to be eccentric and excessively ambitious among his co-workers; however considered a genius.

Nonparallel structures (revision)

You should also be on the look-out for nonparallel structures when revising for correctness and check that there is consistency in verb, person, number and structure. Even though parallel structures are not grammar, strictly speaking, they are definitely on the borderline between grammar and style so a reminder here is worthwhile. Also remember that correlatives like *both...and, either...or, neither...nor, not only...but also* are inherently parallel structures.

> e.g.   (*Nonparallel*) *He used this instrument to measure vibration and this formula has been used to calculate the deviation.
>
> (Two nonparallel structures here: *active simple past tense – passive past perfect tense*).

(*Parallel*) He used this instrument to measure vibration and (he used) this formula to calculate the deviation.

e.g.   (*Nonparallel*) *This mechanism has revolutionized electronics and another achievement of this mechanism is that it has contributed to the development of better safety systems.
(*Parallel*) This mechanism has both revolutionized electronics and contributed to the development of better safety systems.

INDIVIDUAL TASK

4-10 Read the following two passages and identify the nonparallel structures. In order to make the texts more readable and coherent, rewrite the sentences by making them parallel.

What is Biotechnology?

While biotechnology has been designed in many forms, in essence it implies the use of microbial, animal or plant cells or enzymes to synthesise materials, to breakdown materials or to have materials transformed. The aims of the European Federation of Biotechnology (EFB) are: i) to advance biotechnology for the public health; ii) awareness, communication and collaboration promotion in all the fields of biotechnology; iii) to provide governmental and supranational bodies with information and informed opinions on biotechnology and iv) to promote public understanding of biotechnology.
The EFB definition is applicable to 'traditional or old' biotechnology and can also be applied to 'new or modern' biotechnology. 'Traditional' biotechnology refers to the conventional techniques that have been used for many centuries to produce beer, wine, cheese and many other foods. The methods of genetic modification by recombinant DNA and cell fusion techniques, together with the modern development of traditional biotechnology, are referred to as 'new' biotechnology. The difficulties of defining biotechnology and the resulting misunderstandings have led some people to abandon the term 'biotechnology' as too general and instead they have replaced it by the precise term of whatever specific technology was being used.

Adapted from: Smith J. E. (1981). *Biotechnology.* Cambridge University Press: Cambridge (3rd ed.), 2-3. Reprinted with permission of the publisher.

Glass Fibers

Glass fibers are unique materials that exhibit the familiar bulk glass properties of hardness, transparency, being resistant to chemical attack, stability, and inertness, as well as fiber properties of strength, flexibility, are very light, and processability. Glass is made by fusing silicates with soda, potash, with lime, and various metallic oxides.
Two basic processes are used to manufacture continuous glass filaments: marble melt and direct melt. In the marble melt process, glass marbles are first produced by

melting the appropriate mixture of raw materials and forming marbles, which are usually 2 to 3 cm in diameter. And then you remelt these marbles at the same or different location and then one forms the glass fiber product. The raw materials are melted and formed directly into the glass fiber product when you are in the direct melt process. Direct melt is the primary process used in continuous glass fiber manufacture today.

COLLABORATIVE TASK

4-11 Read the following passages and revise them looking for any of the grammatical mistakes seen in this section on grammatical accuracy. For conciseness purposes, also reduce the relative clauses in the passages whenever possible. Then compare your answers with those of your partner.

Passage 1

The history of Sheetco may be useful to illustrate this point (1). This year, the manager of the company, manufacturer of special tools and presses necessary to work with laminated plastic sheets like Formica, have decided to acquire a new installation for the two product lines of the company and improve the customer service department (2). Has become necessary to reduce the company's distribution channels with all these changes (3). In order to improve the first product line, one needs to understand the manufacturing process of the sheets that Sheetco produces (4). First, the laminated plastic sheets are made of specially processed papers impregnated with resins (5); then these sheets, or layers, are cured under intense heat and pressure (6). The layers which are subsequently fused into sheets usually about 1/16 inch thick (7). The end product is sheets that can be used as a surface material, on many types of domestic, commercial and industrial furniture and furnishings, bonded to plywood and hardboard (8). An example of one of the most popular applications of these sheets in the home is to be found in kitchen counter tops (9). As to the second line, Sheetco also manufactures their own bonding and forming presses and sells them, as said above (10). The 'Plus Vacuum' press is based on a heat process, the Company's best press which saves a great deal of time over the conventional cold-pressure method of bonding plastic sheets to a second surface (11). (…)

Passage 2

Many of our modern technologies require materials with unusual combinations of properties that cannot be met by the conventional metal alloys, ceramics, and polymeric materials (1). This is especially true for materials needing for aerospace, underwater, and transportation applications (2). Aircraft engineers are increasingly searching for structural materials that they have strength, stiffness, low densities, and corrosion, abrasion and impact resistance, for example (3). It is a rather formidable combination of characteristics because strong materials are frequently relatively dense increasing the strength or stiffness generally results in a decrease in impact strength (4).

Integrated Circuit Technology

Passage 3

As you know, the study of a circuit building block, the operational amplifier (op amp) is of universal importance (1). Early op amps were constructed from discrete components (vacuum tubes and then transistors and resistors) and its cost was prohibitively high (2). In the mid-1960s the first integrated circuit (IC) op amp was produced (3). An IC op amp is made up of a large number of transistors, resistors, and (sometimes) a capacitor connected in a rather complex circuit (4). Although its characteristics were poor by today's standards and its price was still quite high (5). Its appearance signalled a new era in electronic circuit design (6). Electronics engineers in large quantities started using op amps, which it caused their price to drop dramatically (7). They also demanded better-quality op amps (8). Semiconductor manufacturers responded quickly and within the span of a few years high-quality op amps became available at extremely low prices (9).
In order to take proper advantage of the economics of integrated circuits, designers have had to overcome some serious device limitations (such as poor resistor tolerances) while exploiting device advantages (such as good component matching), therefore an understanding of device characteristics is essential in designing good integrated circuits (10). Also is helpful when applying commercial integrated circuits to system design (11). Germanium and gallium arsenide are also used to make semiconducting devices, however silicon is still the most popular material and will remain so for some time (12). The physical properties of silicon makes it suitable for fabricating active devices with good electrical characteristics (13). In addition, silicon can easily be oxidised to form a layer which is insulating (glass) (14). They have proved that this insulator is used to make capacitor structures and allows the construction of field-controlled devices (15). It also serves as a good mask against foreign impurities, which could diffuse into the high-purity silicon material (16). This masking property allows the formation of integrated circuits; active and passive circuit elements can be built together on the same piece of material (substrate) at the same time and they can be interconnected to form a complete circuit function (17).

4.4 Editing for style

Having checked for correctness and therefore eliminated grammar and usage mistakes, the next step in the post-writing stage is editing for style. Remember in the pre-writing stage we defined style as the way the message is expressed, which is determined by the choice of grammar, syntax, vocabulary and idiomatic expressions. We also described the three main varieties of style that emerge from this choice, that is, *formal, informal* and *slang*. Given that technical and scientific discourse usually moves between the formal and neutral varieties of style, in this section we will focus on those linguistic aspects, within those varieties, that contribute to improving the text and make the most of it.

Writing a text free from grammatical mistakes and that meets the minimum standards of accuracy, punctuation and spelling does not guarantee its adequacy and efficiency. So if you want to generate documents that are more than minimally effective you will have to consider exercising choice; that is, selecting the most appropriate alternative to make the text more readable, informative and persuasive. By checking for style your audience will save time reading the text and will follow your ideas more easily.

The stylistic revisions that you will have to make to improve and polish your writing are the following:

- Checking for sentence length and load
- Controlling the use of passive forms
- Choosing adequate vocabulary
    - ✓ Making an appropriate use of jargon, sub-technical vocabulary and general vocabulary
    - ✓ Controlling redundant and wordy expressions
    - ✓ Avoiding nominalizations and controlling the use of prepositional phrases
    - ✓ Avoiding sexist language

Checking for sentence load and length

A good text should have a balanced combination of short and long sentences. Thus, sentences that are excessively long and overloaded should be avoided since they are too difficult to understand. Readers cannot remember all the information packed in these sentences. Very often they have to reread them because by the time they reach the end of the sentence they have forgotten the beginning.

Example: **Long and overloaded sentence**
Spending more money for repairing the broken mechanism that the technician, who was specially appointed by the director to do the job of revising it, reported as too old would be useless.

Clearer version with shorter sentences
The director appointed a special technician to revise the broken mechanism. The technician reported that the mechanism was too old to be repaired and that spending money on it would be useless.

On the other hand, both short and simple sentences should also be avoided as this creates an unsophisticated and excessively plain style. Firstly, too many short sentences make the text choppy and disconnected. This, we could say, is a common error among non-native writers who do not want to run the risk of writing longer sentences for fear of making mistakes. They prefer to stick to writing short sentences as they can more easily cope with the proper syntactic order. This practice, obviously, results in a poor and somehow childlike style.

Example: Disconnected sentences
The discovery of optical fibre was a major breakthrough in communications. It made more efficient communication systems possible. The optical fibre has a higher cost than coaxial cable. It allows a faster and safer communication transmission. Most communication systems use this material.

Connected sentences
The discovery of optical fibre was a major breakthrough in communications because it made more effective communication systems possible. Although the optical fibre has a higher cost than coaxial cable, it allows a faster and safer communication transmission. Because of this, most communication systems use this material.

Consequently, sentences that are related must be combined to form unified sets of information, in which the relationships between sentences will be openly established. There are two main ways of combining sentences: *coordination* and *subordination*. Even though both ways serve the same purpose, if you want to join sentences and indicate the semantic relationship between them, subordination offers a wider range of possibilities and therefore allows more sentence sophistication. So whenever possible try subordinating sentences and keep coordination for simpler relationships.

Secondly, the unjustified use of sentence beginnings with *There is/are* should also be avoided as they make sentences longer and weaker, and a continuous use of them creates a poor style. These are sentence openers that unskilled writers, especially beginners, are quite fond of given their simplicity and plainness.

(*Weak*)   e.g. There are some problems in the design of the circuit that can be easily solved.
(*Improved*)   Some problems in the design of the circuit can be easily solved.
(*Weak*)   There is a problem of organization in the production plan.
(*Improved*)   The production plant has a problem of organization.

Contrary to some writers' belief, beginning with *There is/are* does not add emphasis to the sentence; instead, it makes the sentence weaker as it does not add any meaning. Besides, in some cases such as the second example above, the suppression of *There is* allows the key words *a problem of organization* to be moved to the end of the sentence, thus giving them more emphasis.

INDIVIDUAL TASKS

4-12 The following paragraphs contain overloaded sentences. Rewrite them providing a shorter and clearer version.

Paragraph 1
We have also learnt that the technical skills of an engineer who has a postgraduate degree from a foreign university that has a very good academic reputation are very useful to the development of his/her professional career if of course this degree focuses on the specific area the company needs additional resources for, which more often than not, is not the case.

Paragraph 2
The first conclusion we reached was that, additionally to the deep and sound knowledge about engineering, the candidate who displayed the most adequate qualities should have some special abilities, namely social and communication skills, which are absolutely necessary for an engineer working in a Production Department, however, the research concluded that the experience carried out is only important as far as it helps candidates to improve the skills mentioned above.

Paragraph 3
Air pollutants such as sulfur oxides, acid gases, volatile organic compounds, nitrogen oxides and carbon monoxide are *gaseous* and air pollutants such as cement dust, smoke, acid mists, metal fumes and fly ash are *particulate*, some of which like smoke are typically easier to see with the naked eye than gaseous emissions.

Paragraph 4
I am writing to let you know that there are problems that I still need to solve to finish my project: the main problem is that I don't know yet what kind of transmission I am going to use and without this information I can't make a good final decision as to the car's dimensions, so this is delaying my work in the project.

Paragraph 5
Living in a dormitory is more uncomfortable but more interesting than living at home. The final decision will depend on the kind of person you are or you want to be and it also gives you a good opportunity to decide whether you want to have the easier way of life, which means continuing living with your parents as well as to think if it is time to learn how difficult life is and what it is to became an adult, mature person.

4-13 Combine the set of sentences below into longer and unified sentences. Use as many connecting expressions as necessary to indicate the semantic relationship you want to establish between them, considering there may be more than one way of combining them. Also avoid too simplistic sentences beginning with *there is* and *there are*.

1. The employee was tired of waiting for a promotion.
    He accepted a new job.
    The new job was about 50 miles from his house.

2. The conference on microelectronics was given by a well-known expert.
    The expert had made important discoveries in this field.
    The expert had been awarded an important prize for his discoveries.
    The conference had a massive attendance.

3. Speech-recognition devices are data entry devices.
    Speech- recognition devices can recognize the words spoken by a person.
    Speech-recognition devices are expensive.
    Speech-recognition devices are not mass-produced.

4. The power was very low.
    The processor could not work at its normal speed.
    We could still manage to finish the work.

5. You must work hard to prepare the job interview.
    You must consider all possible questions about you and the enterprise.
    You may have trouble finding the proper answers.

6. I told Eric to wait in my office.
    I was photocopying the report.
    The phone might ring.

7. The GPRS (General Packet Radio Service) is a wonderful tool.
    The GPRS enables customers to use their phones for services other than voice.
    Customers will be able to access their company's intranet, book flights, check the weather or find out what's on TV.

8. There was a building that showed suspicious scratches.
    They decided to repair it.
    They wanted to sell it.
    They wanted to buy a new one.

9. A research group had collected the latest and most relevant information.
    There was a group of specialists who studied it thoroughly.
    The group of specialists could not reach a satisfying conclusion.

10. A CD is a laser-read data storage device.
In a CD audio, video and textual material can be stored.
A CD stores information in digital.
The conventional tape stores information in analog.
A CD has a complete absence of background noise.

Controlling the use of passive forms

The passive form has certainly proved to be very appropriate in technical writing, as it serves two technical writing purposes; namely, placing the emphasis on the object or event instead of on the subject and not mentioning the agent when it is irrelevant. Yet, it must not be used at random as some technical writers seem to do. If you cannot justify the passive, use the alternative active form. In general, reviewers advise using the active form because it is shorter and more direct.

In technical writing the passive form is justified when:

- You do not know who performed the action
  e.g. About 250 million tires are discarded in USA every year.

- The agent is of no interest to the readers or is redundant
  e.g. In this new edition of the book, some major modifications have been introduced in chapters 2, 3 and 5. (mentioning the agent *the author* is redundant)

- It contributes to giving coherence to the text
  e.g. Operators in the emerging countries face several challenges when it comes to offer broadband/DAL services in their home markets. These challenges can be grouped into three categories: strategic, sales and marketing and operations.

- The sentence has an impersonal meaning
  e.g. Two computers are said to be interconnected if they are able to interchange information.

- You want to be indirect or inoffensive
  e.g. The invoice must be revised before sending it.

- It keeps topic continuity
  e.g. Service robots for personal use are recorded separately as their unit value is only a fraction of that of many types of service robots for professional use. They are produced for a mass market…

4-14 Decide whether the following passive forms are justified or not, taking as reference the previous explanation. Try to justify your choice.

1. XML files are accessed by Linux users everyday to complete their daily work.
2. The strategic devices are not dealt with in this manual.
3. Not much has been done in the last few years to improve the working conditions in the lab.
4. The door should be closed every time you go out.
5. It was agreed by the personnel manager and the union that workers should wear protective glasses while manipulating certain materials.
6. The final grade will be sent by the teacher next week.
7. The importance of telecommunications can be clearly seen in our society.
8. The transistor was invented at Bell Laboratories in 1948.
9. Millions of televisions are produced by Japanese enterprises every year.
10. It has been decided by the board to send the information about the company's stocks every three months instead of every month.

4-15 Study the use of the passive form in the following paragraphs and change passive forms into active ones if they are unjustified.

Radioactive Waste

Radioactive or nuclear waste is the unusual byproduct of nuclear activities. These wastes are classified as high-level waste (HLW), transuranic (TRU) waste, or low-level waste (LLW). Each of these types of nuclear wastes presents its own challenges for proper management.

High-level waste (HLW) is the highly radioactive waste resulting from reprocessing spent fuel, which is the fuel that has been irradiated in a nuclear reactor. Currently, in the U.S., reprocessing involves removing the plutonium and uranium from spent fuel generated by the USDOE nuclear reactors. The plutonium and uranium are recycled for use in defence programs. What remains after reprocessing is highly radioactive waste that must be remotely handled behind heavy protective shielding. Long-term isolation, typically in an underground repository, is required by this waste, while it stabilizes. Shipping HLW presents a unique hazard, so it is packaged in heavily shielded containers for storage and transport.

The spent fuel resulting from the generation of electricity from U.S commercial nuclear power plants is currently not being reprocessed. The Nuclear Waste Policy Act of 1982 is the federal statute that creates the framework for managing nuclear waste. Under this act, the nuclear wastes from nuclear power plants are to be placed in an underground repository for long-term isolation. Ambitious deadlines for siting, constructing, and operating two geologic repositories were set forth by the act. Five years later, the Nuclear Waste Policy Amendments Act of 1987 was passed by the

Congress. These amendments built on the previous direction of the underground repository program and in fact required that DOE focus on one site --Yucca Mountain in Nevada. Two repositories are currently being developed –Waste Isolation Pilot Plant (WIPPs) in New Mexico for waste generated by the DOE, and Yucca Mountain for waste generated by commercial power plants.

Adapted from: Boyce, A. (1997). *Introduction to Environmental Technology.* Vas Nostrand Reinhold, 394-395.

Choosing adequate vocabulary

We saw at the beginning of this book how important vocabulary is in relation to defining other elements of technical writing such as audience, tone and style. More specifically, the expertise of the audience, the level of formality and the tone of the document will determine the vocabulary used. It stands to reason, then, that writers will need to display a good command of the English language, especially at the word and phrase level, to master an effective style. They will need to be aware of the choices they have and the ways in which their choices affect aspects of tone such as pomposity, assertiveness or impersonality and style such as clarity, accuracy, level of technicality, conciseness and readability in general. Because vocabulary is such an important aspect of tone and style, we provide several recommendations that will help you revise the vocabulary used in your writing.

Making an appropriate use of jargon, sub-technical vocabulary and general vocabulary

The level of technicality of a document must be in accordance with its audience. If you think you need to use technical terminology that might be incomprehensible to the audience, you should define it either in the same text or in a glossary. Likewise, when deciding on general vocabulary, you should also make an appropriate choice between more or less formal terminology.

Jargon —the specific technical vocabulary used in a particular discipline, profession or trade—is perfectly appropriate when addressing specialists, but inappropriate for an audience with members from different disciplines as it can become an obstacle to communication. A careful control of technical vocabulary becomes necessary in contexts where the audience includes people from different disciplines. Remember that you should use jargon only to improve your text, not to impress your readers.
The following extract is an example of a piece of writing addressed to specialists where jargon is purposely used. The writer is using jargon because he knows that the reader is familiar with the technical vocabulary used and because it is the shortest, most direct and precise way of transmitting the message.

> A novel time hopping spread-spectrum wireless communication system called Ultra-Wide Bandwith (UWWB) radio is employed to provide low power, high data rate, fade-free, and relatively shadow-free communications in a dense multipath environment. Performance of such communication systems in terms of multiple-access capability is estimated for digital data modulation formats under ideal multiple access channel conditions.
>
> With permission from Moe, Z. et al. ATM-based TH-SSMA network for multimedia PCS. *IEEE Journal on selected areas in communications*, vol.17, no. 5, May 1999, 824. © 1999 IEEE

Sub-technical vocabulary includes terms with the same technical meaning across different disciplines and those that come from general English but take on a different and more specific meaning in a technical context. For example,

- *corrosion* meaning chemical deterioration can be found in different disciplines such as chemistry, materials engineering, mechanics, aircraft engineering, marine engineering, etc.

- s*ound waves, magnetic waves, radio waves, etc.* have a specific meaning which can be traced back from the meaning of *wave* in general English; that is, the word *wave* extends its meaning in more technical contexts, creating new and analogous meanings.

Whether it is jargon or sub-technical vocabulary, the terms you choose must be accurate and precise. Certain objects or concepts must be referred to by their name when there is no other way to refer to them; simplifying, paraphrasing or using a synonym may entail imprecision and confuse the reader. Don't write *an apparatus used for displaying periodical signals* when you mean an *oscilloscope*; that is, be concise and accurate when choosing words since imprecision can hinder comprehension and even change meaning, as can be seen in the following examples:

Don't use	Use
A line of pipes (for a line of connected pipes)	A pipeline
A water tank (for a tank full of water)	A tank of water
Machine (for a specific type of machine)	A lathe/a drill press/a milling machine...
Thing or stuff	Aspect/point /material...
The industry concerned with the extraction, refining and selling of oil	The oil industry

As for non-native writers, the fact that they do not have a very good command of English vocabulary results in an overuse of everyday terms and repetitions. In this case, we must encourage these writers to use a larger vocabulary by incorporating synonyms, antonyms and any other lexical alternatives, which contributes to making the text much richer and more precise. What's more, we must instil in them some concern about applying precision in word choice, and so avoid using generic terms and all-purpose words.

General vocabulary includes the use of words that are commonly used in everyday language and makes the texts more readable and persuasive. In general, the use of formal words among which we find words of Latin, Greek or French origin affects the level of formality and the tone. In certain contexts, these more sophisticated terms can sound pretentious and be ambiguous in others.

Pretentious   When writing, he uses a very prosaic prose

Plain   When writing, he uses a very dull and boring style

In this example, the expression *prosaic prose* sounds pompous if it is addressed to a lay audience who is possibly not familiar with two terms that have been borrowed from a literary context and are being used in an everyday context. In the following example the problem lies in the exact meaning of *prosaic*:

Ambiguous   He is a very prosaic lecturer

Clearer   He is a very boring lecturer

Does the writer mean a dull, boring, unintelligent, slow or thick lecturer? Why use an ambiguous word if you can choose one that has a more specific meaning?

Therefore, when revising, remember to adjust the level of formality: increase the proportion of formal terms in formal documents but resort to more informal words if required by the document or context. Some examples of formal and neutral or more informal terms are:

More formal terms	Neutral /informal words
Undertaking	Task
Necessitate	Need
Notwithstanding	Although
Expenditure	Cost
Preserve	Keep
Inaccuracy	Mistake
Succeeding	Following
Distinct	Different
Implement	Tool
Optimum	Best
Upon	On
Withdraw	Pull out
Reduce	Cut down

As we mentioned in the pre-writing stage, a formal style usually reflects a distance between writer and reader, which may be reduced by incorporating a number of informal words, thus turning to a more informal style. Both styles are perfectly appropriate in their corresponding contexts and none is better than the other. However, Romance students of English may write *The productivity rate of employees has deteriorated because they are not well remunerated* not because they purposefully want to be formal and sound somewhat pompous but because they do not have enough vocabulary to find more informal synonyms to write, for example: *The productivity rate has got worse/ worsened because they are not well paid.* In this particular case, we encourage students to make a balanced use of formal and informal terms and expressions to enrich and dynamize their text and, at the same time, adjust it to their intended audience.

Finally, we will consider another aspect relevant to the choice of accurate vocabulary, namely, *collocations*. Technical writers should be sensitized to and concerned about the vocabulary they choose: words should not only be precise and direct, denoting the exact meaning to be conveyed, but also collocate with the co-occurring terms. Collocation refers to the restrictions on word combinations; that is, which verbs and nouns can be used together (e.g. solve a problem/equation/doubt), which adjectives are used with particular nouns (e.g. major/latest/daring innovation), etc. Knowledge of such collocations is vital for the competent use of language, so a certain familiarity with the usual combinations of words that often appear in technical documents is an important asset for writers. Remember that to help you find the most frequent word combinations you can use some of the online resources available.

Let's do a practical exercise on collocations. Can you think of the words that most often collocate with *to carry out* in technical and scientific registers? Once you've written your own list, contrast this with the one below, which has been taken from an online collocations sampler, in order to spot any similarities and differences.

Collocate	Corpus Freq	Joint Freq	Significance
out	76487	693	25.889966
to	918777	508	16.436840
work	26425	36	5.340737
research	5929	26	4.924963
tasks	756	16	3.971708
duties	783	16	3.970698
tests	2240	13	3.512554
operations	1928	10	3.071013
repairs	485	9	2.975800
job	8852	11	2.917103
further	9373	11	2.893589
improvements	607	8	2.796302
orders	1959	8	2.724750
projects	1614	7	2.554435
students	5878	8	2.517342
surveys	359	6	2.427551
checks	680	6	2.407934
functions	776	6	2.402068

instructions	1278	6	2.371390
exercise	2882	6	2.273368
government	23636	11	2.249850
failed	3817	6	2.216229
decided	5273	6	2.127251
search	2277	5	2.083637
order	10295	7	2.063283
safety	3193	4	1.761019
activities	3333	4	1.750540
installation	330	3	1.703531

COLLABORATIVE TASK

4-16 The text below is addressed to technical experts on telecommunications. Therefore, it includes jargon, sub-technical vocabulary, acronyms, abbreviations and Latin terms. Read it and identify all these terminological aspects typical of expert audiences. Then contrast your answers with those of your partner.

The Detection, Classification and Geo-location of UWB Pulses

The wide bandwidth of UWB pulses precludes the use of traditional receiver architectures. Instead a simple sensor has been employed that detects energy that is instantaneously present over a bandwidth of several GHz.
The sensor fulfils three functions; namely, detection, classification and geo-location. A high gain, spinning antenna is used for detection and will also provide better reception for classification. An IRA has been developed to best meet these requirements. Broadband, omni-directional antennas are used for the three geo-location channels.
The electronics of the sensor consists of a detection system and a broadband collection system. The detector has been designed to detect UWB pulses in the presence of interference and provide a trigger to the collection system. High speed ECL logic has been used to minimize propagation delays through the detector. A Tektronix TDS744404 with an analogue input bandwidth of 4GHz has been used as the collection system. One channel is used to collect data from the IRA and the other three channels are used to collect data from the omni-directional antennas.

Source: Morgan, G. D. et al. The detection, classification and geo-location of UWB pulses. *Microwave Engineering Europe*. March/April 2003, 23. Reprinted with permission of Microwave Engineering Europe, CMP.

INDIVIDUAL TASKS

4-17 The following sentences are intended for a mixed audience or for an audience of laypeople. In order to cater for diverse levels of expertise, rewrite them and make them clearer and to the point with the help of a good dictionary. Remember that you should:
- Avoid jargon
- Define technical terminology, acronyms and abbreviations
- Avoid ambiguous and pompous terms
- Replace formal terms with more informal words

1. Tell the secretary to take an additional copy in case the general manager's oblivious PA fails to remember.
2. The enhancements applied to the girder didn't guarantee its bending strength.
3. The preliminary control yielded important information and established the basis for the subsequent controls.
4. Some of the most popular AI methods used in data mining include neural networks, clustering and decision trees.
5. The client showed an ostensible disagreement as regards the way the transaction was performed.
6. PCBs, circuits and other electrical devices were all placed in large boxes with the purpose of initiating the change of laboratory promptly.
7. If you should purchase a pen drive, make sure it has a minimum capacity of 1Gb.

4-18 The following sentences are intended for a technical expert audience. Now rewrite them in a more formal style with the help of a good dictionary. Remember that you should:

- Use precise terminology
- Avoid unnecessary definitions, exemplifications or paraphrases of technical terminology
- Use acronyms and abbreviations
- Replace informal words with more formal terms

1. The *process of finding and correcting the errors* of the *first* program was a very *difficult job*.
2. Ground-wave propagation is the *most important form* of propagation for frequencies in the *frequency modulation* band. This is the frequency band used for *amplitude modulation* broadcasting and radio broadcasting *for ships*.
3. They *tried hard* to protect the program from being *illegally copied*.
4. The bolt weakened because of *a strong process of deterioration*, causing *the metal rod that transfers movement in the engine* to break.
5. Another major *plus* of *programs for drawing with the help of a computer* is that they can *have connections with* other programs.

6. The *first* step in a radio system is to *choose* the required *radio frequency* wave from those *picked up* by the aerial.
7. I hope you *see the things* that the trainee engineer has proposed to *improve efficiently* the tasks on *the line where pieces are assembled.*
8. With Apple's *small portable music player,* anybody can now *send* a *mountain* of songs into this device at *good* prices.
9. More powerful and faster *personal data agendas* are coming out *right now*. The latest models *have* a *receiver that collects data from a satellite and computes its location.*

Controlling redundant and wordy expressions

Redundant and wordy expressions often result in an inflated style. In order to avoid this excessively pompous style, be on the lookout for these expressions and find more direct, clear and concise substitutes.

Redundancy consists in repeating two words that mean the same unnecessarily. Some of these expressions have become idiomatic, fully incorporated into the English language, and because of this they are used automatically by writers who are unaware that these expressions are a combination of words with the same meaning. Some examples of redundant expressions are:

Redundant	Direct
Join together	Join
Basic fundamentals	Fundamentals
Completely eliminate	Eliminate
Alternative choices	Alternatives
Prove conclusively	Prove

Following is a list with some more redundant expressions; can you complete the list with direct expressions?

Redundant	Direct
Basic essentials	
Consensus of opinion	
Mutual cooperation	
End result	
Utmost perfection	
Local neighbourhood	
Different varieties	
Physical size	
Triangular in shape	
Uniformly consistent	
First introduction	

Redundant	Direct
On a daily basis	
A new innovation	
The month of July	
Actual experience	
Final result	

Wordiness implies saying in three or more words what you can say in one or two. Contrary to some people's belief, the excessive use of longer and more sophisticated expressions does not make your writing more elaborated and profound but rather pompous and pretentious. On the other hand, unskilled writers should bear in mind that the spare use of these expressions is not detrimental to the quality of the text. Some examples of wordy phrases are:

Wordy phrases	Shorter expressions
A great amount of	Many
At the present time	Nowadays
Because of the fact that	Because
First and foremost	First(ly)
At a rapid rate	Rapidly

Could you complete the list below?

Wordy phrases	Shorter expressions
The majority of	
In close proximity	
Readily apparent	
Prior to	
In the course of	
With reference to	
Exhibit the ability	
In the forms of	
A number of	
In accordance with	
In the event that	
With the exception of	
Provide guidance for	
In the near future	
In conjunction with	
Put one in place of another	
To an exceptional degree	

Avoiding nominalizations and controlling the use of prepositional phrases

Another important means of achieving a direct and informative style is to avoid nominalizations and prepositional phrases. If you abuse nominalizations and prepositional phrases you will make your text dense because they lengthen sentences and make the reading slow and cumbersome.

Nominalizations. The excessive use of nouns, mainly abstract nouns ending in *–ion*, *-ity*, *-ent* and *–ness*, makes sentences weak and pompous. Nominalizations generate weak verbs, needless prepositions and sometimes passive forms, which altogether contributes to a less informative style. By replacing some abstract nouns with strong verbs, the sentence becomes more direct and clearer since the emphasis is placed on the action and the agent recovers its original status. Some examples of common nominalizations are:

Nominalizations	Corresponding verb form
provide information	inform
give an explanation	explain
accomplish a modification	modify
conduct an investigation	investigate
create an improvement	improve
make an assessment of alternatives	assess alternatives

Example: **Weak and wordy**
The *maintenance* of the apparatus is better done not only through a periodical *cleaning* but also through the *painting* of the eroded parts and the *replacement* of the worn out pieces.

Direct and clear
The apparatus is better *maintained* if you not only *clean* it periodically but also *paint* the eroded parts and *replace* the worn out pieces.

Prepositional Phrases. A prepositional phrase consists of a preposition followed by a prepositional complement; this prepositional complement is typically a noun phrase or a *wh-*clause or *V-ing* clause. Again, there is nothing wrong in using a prepositional phrase here and there. Prepositional phrases can also be used to write accurately and concisely. As with nominalizations, if you realize that a certain chunk contains too many prepositional phrases in bunches, you can use verbs instead or suppress wordy prepositional phrases.

Example: **Weak and wordy**
This is in reply to your letter of inquiry *with reference to* the possibility of a price discount *in the event of* a 1million- item *order*.

Direct and clear
We are writing to reply to your inquiry *about* the possibility of a (price) discount *if* you *order* 1million items.

In a word, remember that in English verbs are more direct, dynamic and agile than nouns and prepositional phrases, so if when revising you spot too many nominalizations and prepositional phrases, try to suppress those that are wordy, replace some of them with verbs and use coordinated or subordinated sentences instead.

COLLABORATIVE TASK

4-19 In pairs, improve the following sentences and make them more direct and less pompous by eliminating unnecessary nominalizations and prepositional phrases.

1. The solution of the problem was found by applying the Fourier transform.
2. First we have the reception of the signal and then its processing.
3. The use of hydrogen cars would greatly contribute to the reduction of air pollution, especially in big cities.
4. With a laser printer, there is an increase of printing quality.
5. The meeting in the human resources department finished by reaching an agreement between the two parties with reference to the mode of payment.
6. My bafflement was due to my recognition of her deep knowledge of the programming language and of her efficient manner of solving tasks.
7. The steps we followed to launch the new product were first the analysis of the market's needs, then the setting of the objectives and finally the presentation of the product.
8. In spite of the fact that he gained on-the-job training with the help of the internship, he left the company on account of alleged mobbing.
9. The tests and maintenance controls within the safety program of the Maintenance Department are done on a yearly basis.
10. The replacement of the hard disk will entail a cost of approximately EUR 120.
11. The outcome of the assessment work from the Environment Quality Agency is that there is a requirement of an impulse toward acceleration in environmental technology and awareness-raising among the general public.

Avoiding sexist language

In section 4.3 Checking for grammatical accuracy, a word of caution was given as to the non-sexist use of pronouns (*his/her*); yet there are more terms other than pronouns to be catered for. If you want your documents to be inoffensive and unbiased, when polishing, you should also pay attention to the vocabulary and expressions that are sexist or inappropriate. Fair and equal treatment should be given to anybody and any group of people, regardless of their gender. Below are some general tips to take into account to achieve such a neutral treatment:

- For a generic use, use gender-neutral words
- Particularly in jobs, use titles that do not have any gender-marking (*spokesperson, flight attendant*).
- If you know the person you are referring to, use the appropriate gender.
- Using *his/her* all the time is cumbersome and makes sentences lengthy. Instead you may use the following alternatives:

Sexist	*e.g. The engineer may often draw on his humanistic training.*
Revise by using the plural noun or pronoun	*e.g. Engineers may often draw on their humanistic training.*
Revise by using the passive voice	*e.g. Humanistic training may often be drawn on in Engineering.*
Revise by using the infinitive	*e.g. The engineer often needs to draw on humanistic training.*

Here is some vocabulary you should try to avoid as well as some alternative suggestions:

Sexist language	Non-sexist language
he (for a generic use)	he/she, one, they
spokesman	spokesperson
chairman	chairperson; chair
salesman	salesperson
man and mankind	humans; humankind
manpower	workforce; personnel

COLLABORATIVE TASK

4-20 In pairs, read the following sentences, identify the sexist terms and replace them with more appropriate alternatives.

1. Every secretary should have reported to her boss at the end of the day.
2. If a workman spots any malfunctioning in the machinery, he is expected to warn the foreman at once.
3. A policeman requisitioned a cargo of man-made earthenware that was being illegally introduced into the country.
4. The chief doctor in the department is responsible for all doctors, nurses and male nurses. His duties include organizing tasks and controlling work, among other things.

5. Some sparks are thought to have caused the small fire in the assembly plant. A very workmanlike employee witnessed what happened and warned the secretaries. The girls quickly called the firemen and the fire was soon extinguished.

6. If an applicant for the electronics engineer post wants to be seriously considered, he will have to hand in an achievement-based curriculum.

4.5 Proofreading and peer review

Proofreading is the last step of the post-writing stage and also the last step of the overall process of writing. At this point, we will be mainly concerned with checking for more mechanical aspects such as format, typography, spelling and punctuation. Needless to say, checking for all these aspects will take some time and maybe more than one reading; however, it can be worthwhile in terms of giving accuracy and credibility to your writing. A piece of writing that is full of typographical, spelling and punctuation mistakes may not only give the impression of having been written carelessly but also, and most importantly, may lead readers to mistrust you and question the quality of your technical work.

To ensure that your writing is free from mistakes, we have developed a list of proofreading tips. The items in this list have been organized in three sections according to typography, spelling and punctuation:

When revising typographical mistakes, check that
- figures, equations, calculations and numerical expressions are correct. Double check any number or calculation; it's very easy to write 4500 instead of 45000, for example.
- titles and subtitles as well as font style and size are consistent. Remember that the main words in titles are written in capital letters.
- abbreviations and acronyms are properly used. Remember that acronyms are written in capital letters. Also remember to write the full form of an acronym the first time you mention it.
- the layout is consistent. This includes all aspects concerning appearance such as margins, borders, rulers, indentation, pagination, justification and spacing.
- the graphical information is accurate. Check how your graphics are presented, how their captions are written, how figures are numbered and how graphs are annotated.

When revising spelling mistakes, use
- spell checkers to revise your spelling mistakes. Use them with caution, however. Spell checkers do not spot every single mistake. For example, they might miss proper names, incorrectly used words (practise and practice) and technical terms.
- a good dictionary. Mastering English spelling is not an easy task even to native speakers. Many English words are written quite differently from the way they are pronounced.

When revising punctuation mistakes, check that
- punctuation marks have been correctly used. If you feel that you do not have a good command of basic punctuation rules, you should consider using a manual of style or a book on writing, where you will most certainly find a full list of punctuation rules. In the handbook at the end of this book you will also find a summary of the main punctuation marks and their use.

In order to carry out a thorough revision of your document with regard to not only mechanical aspects but also all the other aspects that intervene in the writing process, here are some techniques that can contribute to detecting trouble spots:

- *Ask someone else to read it.* Do practice what is known as *peer review*. A peer will be able to catch errors that we, as writers of the text, miss or overlook because we sometimes read what we think we wrote, not what we actually wrote.

- *Put the writing aside.* Don't proofread right after writing; it is a good idea to leave the text to rest for some time in order to be more effective.

- *Check one point at a time.* You should check for one or two aspects at a time, especially with long texts. If you try checking everything at once, you will most probably overlook some errors.

- *Read it aloud.* By reading the text aloud we can hear errors that our eye could not see. For example, a long and awkward sentence is very easily detected when reading aloud.

- *Read backwards.* This is a technique to basically catch typographical errors as the reader is forced to read word by word.

The use of one technique does not exclude the use of the others, on the contrary; they are often used in combination. For example, you may use the technique *check one point at a time* first to revise sentence structure, grammar and word choice and then the *read backwards* technique to spot typographical errors.

COLLABORATIVE TASK

4-21 First, individually write a paragraph using one of the patterns of paragraph development described in this book. You may choose any technical or scientific topic you like and know well. Then exchange paragraphs with a peer and revise each other's paragraph with the help of the questions below. Finally, give your colleague some feedback on aspects that could be improved and assign up to one point to every question in order to provide a final mark to the activity. Note that question ten is a general question and that you should assess the paragraph as a whole.

Questions for peer review:

1. Is the topic or main idea of the paragraph clear?
2. Can you identify any topic sentence? Is it appropriate?
3. Are all sentences related to the topic sentence? If not, which sentences are off topic?
4. Are all the supporting ideas included? Is the amount of detail adequate?
5. Does the paragraph display a consistent pattern of organization? Are ideas logically and clearly organized?
6. Are sentences properly connected?
7. Is the vocabulary used adjusted to the intended audience?
8. Is there any sentence whose meaning is not clearly expressed?
9. What are the main grammatical, stylistic and spelling mistakes?
10. What do you think are the strengths and weaknesses of this paragraph? How would you improve these weaknesses?

Question	Feedback	Grade
1		
2		
3		
4		
5		
6		
7		
8		
9		
10		
	TOTAL	/10

INDIVIDUAL TASK

4-22 In this concluding task you will reflect on your writing performance in terms of the four basic aspects we have dealt with in the post-writing stage. That is, you will have to identify with which of the following aspects you usually have problems and how often:

- **Content and organization** (thematic unity, patterns of organization, adequate amount of detail, paragraph development, coherence)
- **Grammatical accuracy** (subject-verb agreement, pronoun agreement, pronoun antecedent, subject repetition/omission, dangling modifiers, reduced relative clauses, run-on sentences, sentence fragments, parallel structures)
- **Style** (sentence length and load; the use of passive forms; the use of jargon, sub-technical vocabulary and general vocabulary; redundancy and wordiness; nominalizations and prepositional phrases; and sexist language)
- **Proofreading** (spelling and punctuation)

In the table below, fill in the frequency columns with the aspects you have problems with.

	Rarely	Often	Very often
Content and organization			
Grammatical accuracy			
Style			
Proofreading			

4.6 Academic and technical sample texts

In this section you'll find three different types of documents: an academic essay, a lab report and a progress report. The activities associated with each text will make you reflect on all that you've learnt about the writing process and will give you a general overview of how this process can be applied to both academic and technical texts.

Revising an academic essay

Below there is a sample of an essay written by a student. The drafts correspond to different revisions within the post-writing stage. In the first draft only content and organizational aspects were considered whereas in the second the focus was on grammatical accuracy and style.

First draft. In this first draft, the student revised only content and inappropriate or wrong organizational and structural aspects following the teacher's indications.

Hydrogen (Improve title)

The way of life of developped countries and the industrialization in emerging countries such as China and India is causing that fossil fuel supplies run out. The need to find a new combustible, which keep survival our style of life can be satisfied thanks to the use of hydrogen.

This new energy source can be a great solution given that it is an endless and clean resource with a great variety of uses. (Revise *introductory paragraph and thesis statement*)

In principle, a hydrogen cell operates like a battery: it produces energy in the form of electricity and <u>heat</u> (Revise *content*); as long as fuel is supplied. A fuel cell consists of two electrodes sandwiched around an electrolyte. Oxygen passes over one electrode and hydrogen over the other. As long as the fuel and oxygen is supplied to the cell, this will keep producing electricity for ever. The oxygen needed by a fuel cell is usually simply obtained from air. Hydrogen cells can be thought of as devices that do the reverse operation of that in the well-known experiment where passing an electric current through water splits it up into hydrogen and oxygen. They can be used to produce quite small amounts of electric power for devices such as portable computers or radio transmitters, right up to very high power ratings for electric power stations.

In fact, (Topic *sentence missing*) not only hydrogen is used in cars but also in airplanes and even in some household appliances like electric scales. The development of the cell is focused in cars since they are the main future market. Some of the giant car companies are also desinging hydrogen-powered cars, for example BMW is going to sell in 2007 one hundred cars whose fuel is not petrol but hydrogen. Furthermore, British Petroleum (BP) has opened a hydrogen station in Beijing in order to encourage people to use hydrogen-powered cars. Finally, governments are also financing projects that investigate how to make a better use of hydrogen cells and are sponsoring initiatives that involve the use of hydrogen. This way, cars which use renewable sources will be free of taxes from now on.

(Improve inter-coherence) A point interesting to be mentioned are the differences between petrol and hydrogen cars. On the one hand, petrol cars have both a better mecanical efficiency and are also powerful. On the other hand, hydrogen is an endless source of energy, it does not produce polluting emissions, whereas it damages the ozone layer. *(Revise content)* Despite hydrogen cars are less powerful than fuel combustible ones, BMW has managed to manufacture a car with 285 HP that accelerates from 0 to 100 Km/h in six seconds and reaches a maximum velocity of 302 Km/h. It is a great example of how the latest projects can give benefits to companies. *(Good concluding sentence)*

In conclusion, we could say that hydrogen is a good solution to resolve the shortage of fossil fuel sources and contamination. Because of people obtain hydrogen from water in nature, an abundant resource that it is longer than human life, and making a simple chemical process.

We engineers are able to obtain a clean energy source which can be used in different areas, mainly in cars. Hydrogen is the future. *(Revise concluding paragraph)*

Second draft. Drawing on her teacher's indications, the student submitted this second draft, which incorporates changes in terms of paragraph division and development. Now a grammatical and stylistic revision was necessary, so the teacher suggested where and what kind of mistake had been made.

Hydrogen: the future is coming

The way of life of developped *(Idiomatic expression and spelling)* countries and the industrialization in emerging countries such as China and India is *(Subject–verb agreement)* causing that fossil fuel supplies run out. *(Syntax)* The need to find a new combustible, which keep *(Subject–verb agreement)* survival our style of life *(Idiomatic expression)* can be satisfied thanks to the use of hydrogen. This new energy source can be a great solution given that it is an endless and clean resource with a great variety of uses. First we are gonna *(Style)* compare hydrogen with fuel cells and then we will see the applications and the advantages of the hydrogen cell. *(Good thesis statement)*

In principle, a hydrogen cell operates like a battery: it produces energy in the form of electricity and, as a result of some undesired losses, also heat; as long as fuel is supplied. *(Sentence fragment)* A fuel cell consists of two electrodes sandwiched around an electrolyte. Oxygen passes over one electrode and hydrogen over the other. As long as the fuel and oxygen is *(Subject-verb agreement)* supplied to the cell, this will keep producing electricity for ever. The oxygen needed by a fuel cell is usually simply obtained from air. Hydrogen cells can be thought of as devices that do the reverse operation of that in the well-known experiment where passing an electric current *(Syntax)* through water splits it up into hydrogen and oxygen. They can be used to produce quite small amounts of electric power for devices such as portable computers or radio transmitters, right up *(Style)* to very high power ratings for electric power stations.

In fact, the hydrogen cell has a lot of applications. Not only hydrogen is used *(Word order)* in cars but also in airplanes and even in some household appliances like electric scales. *(Improve intra-coherence)* The development of the cell is focused in *(Preposition)* cars since they are the main future market. Some of the giant car companies are also desinging *(Spelling)* hydrogen-powered cars, for example BMW is going to sell in 2007 one hundred cars whose fuel is not petrol but hydrogen. *(Run-on sentence)* Furthermore, British Petroleum (BP) has opened a hydrogen station in Beijing in order to encourage people to use hydrogen-powered cars. Finally, governments are also financing projects that investigate how to make a better use of hydrogen cells and are sponsoring initiatives that involve the use of hydrogen. This way, cars which use renewable sources will be free of taxes *(Use a compound)* from now on.

Another point interesting *(Word order)* to be mentioned are *(Subject-verb agreement)* the differences between petrol and hydrogen cars in order to better understand the advantages of one over the other. On the one hand, petrol cars have both a better mecanical *(Spelling)* efficiency and are also powerful. *(Non-parallel)* On the other hand, hydrogen is an endless source of energy, it does not produce polluting emissions, and so does not damage the ozone layer. *(Run-on sentence)* Despite *(Connecting expression)* hydrogen cars are less powerful than fuel combustible ones, BMW has managed to manufacture a car with 285 HP that accelerates from 0 to 100 Km/h in six seconds and reaches a maximum velocity of 302 Km/h. It is a great example of how the latest projects can give benefits to companies.

In conclusion, we could say that hydrogen is a good solution to resolve the shortage of fossil fuel sources and ↓ contamination *(Verb missing)*. Because of people obtain hydrogen from water in nature, an abundant resource that it is longer than human life, and making a simple chemical process. *(Rewrite)* We engineers are able to obtain a clean energy source which can be used in different areas, mainly in cars *(Improve vocabulary)*. If we want to respect our environment, hydrogen is the future. At least, this will be so provided water is free and available to most inhabitants in the planet.

Bridging the gap between academic and technical texts

The reports below were also written by students. Compare them with the academic essay above in order to find similarities in terms of information organization and paragraph development.

Lab Session on Mechanics

Definitions: mortar

In order to carry out this lab experiment, an essential substance will be required: mortar. *Mortar* is a type of material which is capable of joining fragments which are derived from one or several substances. In this case, the substances to be joined are concrete and two other less important materials. *Concrete* is composed of different materials, such as cement, pebble gravel, sand, ceramics and small stones. The mechanical characteristics of mortar depend both on the materials that are used and their respective composition. By *composition* we mean the proportion or percentage in which each element contributes to the final mixture.

Objective of the trial

The objective of this lab session is to discover mortar tensile strength. To obtain the results, a trial of mechanical breakage is carried out until the piece of mortar we are working with breaks. The fracture will only take place in the area of the body subject to tensile strength, since the resistance it offers to tensile strength is very low in relation to its compressive strength. Thus, our final objective will be the examination and study of the fissures or, if it is the case, of its total crack or fracture.

Trial description

Firstly, some samples of mortar will be needed before starting the experiment. These samples will have to be handled softly as, despite the fact that they are hard and resistant, they would break if they fell on the floor. Secondly, we will need to verify the dimensions of the piece of mortar, which should have an ideal shape (5x5x20 cm). Thirdly, we will introduce the aforementioned piece into a universal testing machine, where certain forces will be applied. The experiment will be carried out considering the UNE 7-474-92 code.

The plate of the machine will apply a P force at a velocity which will increase 5 Kgf per second, until the mortar breaks. The force of this fracture will be registered on a paper. It is essential to ensure that the faces of the sample be completely smooth.

How to manufacture the samples

The samples of the mortar will be manufactured in groups of three in metallic moulds. The mortar will be made in a cement mixer during two minutes. To be precise, the procedure to be followed with regard to the mixture will be as described below:
- first, 500 grams of cement will be required
- next, 250 grams of water will be poured to the mixer
- finally, 1500 grams of sand will be slowly added

Lab Session on Mechanics (cont.)

Having done that, the moulds will be then shaken to obtain a uniform mixture. It is really important to know that the mortar can only be extracted from the mould 24 hours later. Similarly, the moulds should be at a temperature of 21°C with a relative humidity over 90%.

Procedure of the experiment and results

At this stage, we know that the characteristics of concrete depend on those of mortar. Therefore, taking this into account, it can be inferred that the results obtained in this experiment are the same as those for concrete. To be precise, the trial of force has to be done with cross section samples of 4x4 centimetres and a distance among supports of 10.67 centimetres. The strength can be obtained from Navier's formula. This strength is directly correlated with flexural moments and fracture strength. Thus, to sum up, taking into consideration the characteristics of the machine and Navier's formula, we will be able to obtain the following formula:

$$F_{ct,fl} = 0,25 P_{rot} \, (kgf/cm^2)$$

To conclude, the experiment has allowed us to test mortar tensile strength and the results yielded show that mortar tensile strength directly correlates with flexural moments and fracture strength.

A progress report

Introduction

This report informs about the progress made in my security research project for KLM over the last few weeks. After writing the last progress report a month ago, much work has been done. The outcomes of the research I have done during this past month shed new light on the problem, thereby making me change my initial scheduling. In the next section, the status and progress of the project are summarized; the following section explains the work that still remains to be completed and finally in the last section some questions are raised.

Work completed

Last month I spent most of my time interviewing people. These interviews provided me with unexpected information. First, Mr Smith (head of KLM security department) was interviewed. The main outcomes of this interview were that, in his opinion, KLM security was not strict, or rather, was not strict enough. This baffling news shed new light on my research question '*how can security costs be reduced?*'. Second, Mr Goodman (head of Vueling security) was also interviewed. He gave me a brief explanation of the security systems within his company, which provided me with a benchmark for KLM security. From his interview it became clear to me that KLM security was far from being strict and not that well-organized as one might expect. For me, this posed a problem because it was clear that I should not only focus on the economic aspect but also on the quality of KLM security.

A progress report (cont.)

Work to be completed

After the interviews I held, my schedule and planning had to be revised. Taking into account the information gathered in the interviews, it was then quite certain that the security problem was more serious than what I had initially expected. It was not only an expensive security system but also an ineffective one. In order to solve this unexpected problem, I decided I had to rewrite my research thesis and also to repeat an important part of my research, this time focusing on the efficiency and quality as well. Because of this, I will unfortunately fall behind my initial schedule. However, I am now convinced that I have to introduce these changes to properly address the problem. As a result, I will restart as soon as possible with my literature review and in the meantime will do follow-up interviews with the main people involved. I estimate that this move represents a delay of one month for my project.

Questions

As I said, the new scenario is that reconsidering the research question and starting again gathering information about related literature in the field are certainly going to set back my project. In the light of this, I have three questions: First, do you share my opinion that a full revision of my research question is absolutely necessary? Second, is there any additional literature you can recommend me? And third, is it a problem if my project takes longer than expected? I hope you can answer these questions as soon as possible.

At first glance several similarities between the academic essay and the reports can be easily identified. To begin with, just as the academic essay has an introduction, a body and a conclusion so the reports above also display a similar structure. As in most reports, they contain an introduction, constituted by a definition and the objectives; a body, which includes a description and procedure or a discussion of the work completed or scheduled; and a conclusion with a summary of the results or problems and questions that arise from the discussion. In terms of paragraph development, we also find some sentences acting as a topic sentence or as a thesis statement. The latter usually coincides with the report purpose statement and/or a plan-of-development sentence usually found at the end of the introduction. Besides, some of the paragraphs are organized following one of the patterns of paragraph organization commonly used in academic essays, such as listing, problem-solution or chronological order.

PROJECT

We have reached the final stage of the writing process: the post-writing stage. Now it is time to polish your document to obtain a satisfying and effective final version. This stage is one of the most time-consuming as you will have to revise your document several times in order to spot any inconsistency. Also note that this final stage is the same for the bottom-up and top-down approach as for both you will have to revise your last draft.

Bottom-up approach ⬆ **and top-down approach** ⬇

STEP 4

In this final stage of the writing process you will have to go through four substages or phases to properly finish your document. Basically you will have to:

- *Revise content and organization.* Carry out a revision in two separate readings, one for content and one for organization. With respect to content concentrate on the amount of information included (both central information and details), the proportionality between the importance of information and the length of the explanation, the completeness of the document, the thematic unity of the different paragraphs and the appropriateness of the title. As for organization, check the logic of the document's organization, the paragraph size and distribution, the organizational divisions used and the function of the introduction and conclusion as complete and distinctive sections.

- *Check for grammatical accuracy.* Reread the whole document concentrating on grammatical accuracy in order to make sure that your text is grammatically correct. Eliminate any grammar mistake you may find.

- *Edit for style.* Make sure you have selected the most appropriate style in order to make the text more readable, informative and persuasive for your audience. You should basically revise sentence length and load, the use of passive forms and the choice of adequate vocabulary (the appropriate use of jargon, sub-technical vocabulary and general vocabulary; the control of redundant and wordy expressions; the use of nominalizations and prepositional phrases; and the avoidance of sexist language).

- *Proofread and practise peer review.* Finally you should check some more mechanical aspects such as format, typography, spelling and punctuation. This last revision may take you more than one reading but it is necessary if you want to give credibility to your document.

At the end of this post-writing stage, you should have a complete, accurate and effective document. You can compare this final version to the first draft you wrote in order to see how much it has improved!

ESSAY ASSESSMENT SHEET

This essay assessment sheet can be used both by teachers and students as a guide to assess the three different parts of the project carried out so far. Teachers can use this table after each submission to evaluate students' work and to provide some feedback whereas students can take it as a reference to know about the criteria adopted to assess their work.

Stages	Comments	Assessment
Pre-writing 1. Adequacy of topic according to audience and purpose. 2. Adequacy of style and tone adopted. 3. Idea-generation and gathering of information (citing sources) 4. Outlining		
Writing (drafting) 1. Paragraph development 2. Intra-paragraph coherence 3. Essay development 4. Inter-paragraph coherence		
Post-writing 1. Revision of content and organization. 2. Grammatical accuracy 3. Editing for style 4. Proofreading (spelling, punctuation, others)		
Maturity (Linguistic proficiency and ideas)		
FINAL MARK		

PART III

CHAPTER 5
Grammar, style and punctuation

Introduction to the handbook

As we mentioned at the beginning of the book our main purpose is that you become acquainted with certain types of documents, known as *genres*, and also that you become competent enough and with sufficient writing skills to adapt your documents to every writing situation. However, we know that engineering students come from heterogeneous backgrounds and exhibit different language competence levels. In the light of this, the final part of this book has been designed as a tool to help you improve your competence with respect to different aspects: main constituents in language, inter and intra-paragraph coherence, grammar, style and punctuation. The tasks included to practise these aspects have purposely been designed for individual work, so you can choose those tasks and materials that best suit your needs, English competence and learning style. In this sense, we encourage you to learn effectively and individually and to become responsible for your own learning process.

Apart from the practical tasks included in this final chapter, remember that there are other ways to improve your writing competence. As you probably know, the Internet is a powerful tool for language learning as it fosters the use of authentic communication in English, basically through the integration of reading and writing skills. Some Internet resources you can use when designing your own learning route are:

- *Web-based learning materials.* Some of their advantages are that they are highly interactive, provide immediate feedback (often with explanations) and contain resources classified according to criteria such as level of difficulty or type of task. Particularly useful are placement tests, which will provide you with information about your actual level of English and will allow you to identify your strong and weak points.

- *Search engines.* They may help you learn more about specific expressions and terms by exploring how they are used in authentic texts posted on the web. In this sense, search engines may be used as "dictionaries" that provide you with actual examples of language use.

- *Online concordancers.* By entering a word, phrase or structure, they provide you with a list of examples of terms within their immediate context from a corpus or collection of texts gathered for language study.

- *Online collocation samplers.* They can be used to retrieve a word's most significant collocates from a corpus.

These are just some examples of the online resources that foster an exploratory approach to language and that encourage you to be a more reflective and informed learner as they require you to evaluate your own performance, find answers for yourselves and contrast language forms and texts. The Internet may, in this sense, be particularly helpful if you want to assume a more active role in your learning process. By working autonomously and making informed decisions (concerning the choice of activities, self-assessment, etc.) you will become a more effective learner and thus improve your writing skills and competence.

CHAPTER 5

Grammar, style and punctuation

5.1 Introduction
5.2 Main constituents in language: the phrase, the clause and the sentence
5.3 Revision of intra- and inter-coherence
5.4 Revision of grammar and style
5.5 Punctuation

Reflecting on…

What is most important for you when writing in English? What are your priorities, just to communicate or to communicate accurately?

What kinds of skills (listening, speaking, writing and reading) are you particularly good or bad at?

What are the main difficulties you encounter when writing in English? Can you recall the most common grammatical and stylistic mistakes you make?

Does your motivation to learn technical English writing arise from your professional or academic development, or from a personal interest?

What sort of language learner would you say you are? First choose the adjectives that best describe you (analytical, intuitive, careless, motivated, perfectionist, autonomous, impulsive, cooperative). Now reflect on the advantages and disadvantages of the learning style you most identify with.

What are your strengths and weaknesses as an English language learner?

5.1 Introduction

This last chapter is devoted to providing further practice to those students who feel they need some extra exercises and want to work autonomously to improve their written communication in English. The exercises are intended to help students reinforce linguistic aspects that can be troublesome. The chapter opens with a summarized description of the main linguistic elements of the text, that is, the sentence, the clause and the phrase. Given that these linguistic elements are constantly referred to throughout the course, both in the theoretical and in the practical parts, we thought it would be very useful to include a definition of each, together with some clarifying examples. In general, students at technical schools are very prone to forgetting grammatical concepts they once learned at primary or secondary school, as the main emphasis in technical studies is basically placed on numerical material rather than on any other type of written communication. As a result, we are setting forth a practical revision of intra- and inter-coherence, and grammatical and stylistic aspects, a revision developed through a rich variety of exercises. Finally, another important point is punctuation—often disregarded by technical students—which has been included with the purpose of helping you write more clearly and punctuate appropriately.

5.2 Main constituents in language: the phrase, the clause and the sentence

Before beginning with a practical revision of the aspects mentioned above, and in order not to clutter you up with a long and tedious description of grammatical concepts, we will be giving a cursory glance at the hierarchical structure of language. This will provide you with a sufficiently general picture of the different levels in language hierarchy, essential for you to understand the grammatical and linguistic concepts dealt with in this book. Another advantage of looking at the main constituents in language is that you will become familiar with the most basic terminology and associated concepts. Good writing certainly calls for a clear understanding of the sentence and of at least the next two lower levels that constitute a sentence (i.e., the clause and the phrase). We believe that basic grammatical definitions will enable you to improve your command of the English language.

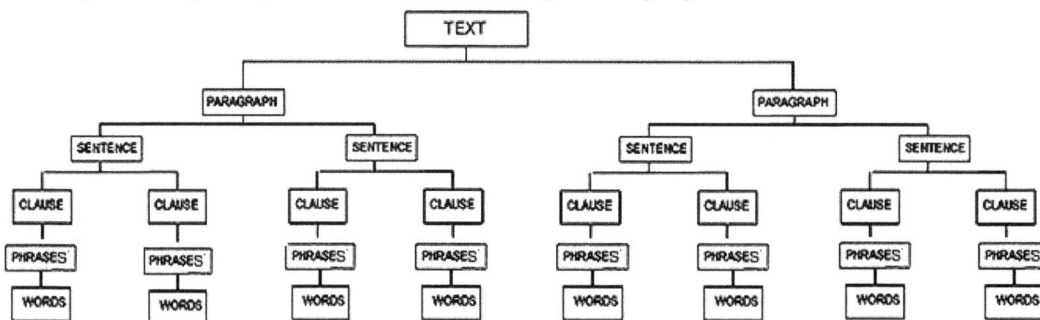

Fig. 5.1 Main constituents in language

The phrase

A *phrase* is a linguistic element made up of one or more than one word and has no subject-predicate structure. In a scale of complexity, it is found between the word and the clause. The main types of phrases are listed below:

- Noun phrase
  e.g. The resistor, the flat resistor, the old flat resistor
- Verb phrase
  e.g. corrode, has corroded, to avoid corrosion, are seriously corroding
- Prepositional phrase
  e.g. on the surface, through the pipe, before the test
- Adverbial phrase
  e.g. yesterday morning, slowly, there

The clause

A *clause* is a group of words comprising a subject and a predicate (Verb, Complement, Object and Adverbial) that constitutes a sentence or part of a sentence. Clauses can be *dependent* or *subordinate* and *independent* or *main*. In the sentence *He went to the lab because he had to use the oscilloscope, He went to the lab* is a main clause whereas *because he had to use the oscilloscope* is a dependent clause. Note that a *main clause* makes sense on its own and therefore can be equated with a sentence. On the other hand, a *dependent* clause cannot stand by itself since it depends on some other element in the sentence and cannot be punctuated as an independent clause. This can be seen in the following examples:

e.g. Although he graduated at the age of 26.

As soon as the gas enters the engine.

Since the mechanism failed.

So that the mistake is not repeated.

Which didn't come out successfully.

Clauses can also be *finite* and *non-finite*. Nearly all independent sentences are finite as they contain verbs that can be used with a subject to make a verb tense (i.e. the verb is inflected for tense, person and number). For example, both *He graduated in engineering* and *Since he is still writing the report*, are finite clauses. On the other hand, *non-finite clauses* are clauses that cannot be used with a subject to make a verb tense. There are four main types of non-finite clauses:

- Present participle clause, e.g. *Leaving the library*, the student dropped his ID
- Past participle clause, e.g. *Covered with sand*, the thermometer was found working.
- Infinitive with *to* clause, e.g. The next step is *to assemble the fittings*.
- Infinitive without *to* clause, e.g. Rather than *copy the document*, he faxed it.

The sentence

A *sentence* is a group of words comprising a subject and a predicate that can stand alone. It can be used to make a statement, give a command, ask a question, etc. A sentence can have just a subject and a verb or more than a clause; that is, it can have a main clause and several dependent clauses.

e.g. The component failed.

Sentence

e.g. Since he used a component which was defective, the engine couldn't start.

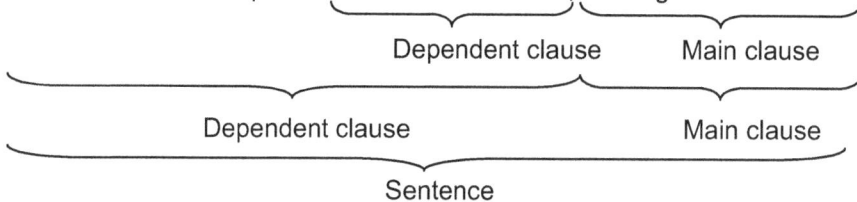

Dependent clause   Main clause

Dependent clause   Main clause

Sentence

Types of sentences

Having seen what a sentence is, we can now look at the four types of sentences:

1) simple sentence
2) compound sentence
3) complex sentence
4) compound-complex sentence

1) Simple sentence:
It contains one independent clause that stands alone. It expresses only one finished thought or single idea. For example:

e.g. I study engineering.

e.g. Different models of electric cars are being constantly improved and marketed.

2) Compound Sentence:
It contains two or more independent clauses that present one complete idea each. In a compound sentence the ideas in each of the clauses are given equal emphasis because the writer considers that the ideas in each of the clauses are of equal importance. There are several ways to join compound sentences; some are joined by a coordinating conjunction or a correlative, some by a sentence connector, and others by a semicolon.

Coordinating conjunctions:

> Independent clause (comma) + coordinator + independent clause

Coordinating conjunctions are: *for, and, nor, but, or, yet, so* (to remember these conjunctions you may memorize the word FANBOYS) and are usually punctuated with a comma.

e.g. They switched off the power, so the information was lost.

At first he didn't know how to create web pages, but he learned quickly.

They have developed a new composite material, and it has proved to work much better.

Sentence connectors:

> Independent clause + (semicolon or period) + (sentence connector) + independent clause

e.g. The mountain bike was protected with a special paint; in addition, it was carefully wrapped with foam packaging.

China is demanding more and more aluminium. As a consequence, the price of aluminium is increasing world-wide.

Operators in this workshop had better wear special clothing; otherwise, they may get hurt.

Readers need to follow your reasoning; they expect to see how exactly you support your claims.

3) Complex Sentence:

This expresses a relationship of dependence. It contains one independent clause which can stand on its own and one or more dependent clauses. Complex sentences use subordinating conjunctions to signal the relationship between the independent and dependent clauses. The order in which the clauses appear is irrelevant, but if the dependent clause comes first, a comma is necessary whereas if the dependent clause comes last, the comma is optional.

Subordinating conjunctions:

> Subordinating conjunction + dependent clause + comma + independent clause

e.g. Since the technician couldn't operate the milling machine, he read the instruction manual.

e.g. Despite being initially enthusiastic, they eventually dropped the course.

In order to upgrade the model, the R+D department has devised several improvements.

> Independent clause + (comma) + subordinating conjunction + dependent clause

e.g. He read the instructions manual (,) since the technician couldn't operate the milling machine.

They eventually dropped the course (,) despite being initially enthusiastic.

The R+D department has devised several improvements (,) in order to upgrade the model.

4) Compound-Complex Sentence:

It is a combination of the compound and complex sentence types. Below are several examples with different combinations:

> Subordinating conjunction + dependent clause + second dependent clause + comma + independent clause:

e.g. As soon as Tim read the email that the manager had sent him, he mentally outlined his reply.

> Independent clause + semicolon/period + sentence connector + second independent clause + subordinating conjunction + dependent clause

e.g. Diesel engines are slower, noisier and more pollutant than petrol engines; however, they are still in use today because they are very economical.

> Independent clause + (comma) + coordinating conjunction + second independent clause + subordinating conjunction + dependent clause:

e.g. Diesel engines are slower, noisier and more pollutant than petrol engines, but they are still in use today because they are really more economical.

Sentence variety

Finally, remember to make a varied use of sentences when you write as it will make your writing richer and, therefore, more interesting. Passages containing only simple or compound sentences are boring and reflect a poor command of the writer's language proficiency. On the contrary, abusing complex or compound-complex sentences will result in inflated and complicated writing. As mentioned previously, try to find the happy medium and apply the variety criterion. One last reason that justifies sentence variety is that you can also use it for a specific purpose. For instance, a short simple sentence is a very effective way to emphasize an idea or point you want to drive home.

INDIVIDUAL TASKS

5-1 Considering that a sentence must have a subject and a verb, read the following sentences and say whether they are Complete (C) or Incomplete (I). Then make the required changes to correct incomplete sentences.

1. Is impossible to determine the exact composition of this product.
2. In the site excavation, found ancient wooden carved sticks.
3. Engineers are said to have a pragmatic, down-to-earth attitude to life.
4. Along the assembly line in the plant, the chassis is painted.
5. Among secondary school students, engineering is regarded as difficult studies.
6. Although they will withstand such temperatures, the concrete beams unsafe.
7. Offers improved safety systems, higher speed and lower gas consumption.
8. But is considered to be a good option.

5-2 Read the following sentences and indicate which clauses are main or independent clauses (I) and which are dependent or subordinate clauses (D).

1. He didn't replace the batteries of the cordless drill; in fact, he went on using it for hours.
2. The batteries of the cordless drill had been recently replaced, so he went on using it for hours.
3. The batteries of the cordless drill needed being replaced. However, he went on using it for hours.
4. The expensive and long-life batteries of the new Japanese cordless drill were flat.
5. The batteries of the cordless drill which he had bought a year ago had been recently replaced; thus, he went on using it for hours.
6. You can use the cordless drill for as many as 7 hours provided its batteries have been recently recharged.
7. Ignoring that even a cordless drill needs some maintenance is neither advisable nor wise.
8. You had better replace the batteries of this cordless drill because they have a 5-hour life.

5-3 Read the following paragraph and identify the types of sentences used:

> When designing an engine, one of the first decisions to take regards the selection of the type of cycle (1). Sometimes, personal preference and previous experience have a great deal to do with the selection of type; however, sometimes there are certain cases where the selection is predetermined by general conditions (2). For example, a small oil engine, for which the fuel economy is not so important as simplicity, should be built on the two-stroke cycle with crankcase scavenging (3). On the other hand, lightweight, high-speed gasoline engines usually have a four-stroke engine, whereas medium-sized gas and oil engines are built to operate on either cycle (4). The decision of a two- or four-stroke engine is often not easy (5).
>
> Adapted from: Maleev, V. L. *Internal Combustion Engines. (1993)*. Singapore: McGraw-Hill Book Company, 306. Reprinted with permission of the McGraw-Hill Companies.

CRITICAL THINKING

5-4 In terms of sentence variety, how does the writer finish the paragraph in the task above? What effect has been created?

5.3 Revision of intra- and inter-coherence

In this section you will find additional exercises to practise different intra- and inter-coherence aspects. If you want to revise the corresponding theoretical explanations you can refer back to sections 3.5 and 3.8 in Chapter 3.

INDIVIDUAL TASKS

5-5 Read the examples below and choose a suitable connecting expression for each sentence. Only one answer is correct.

1. Large firms have made huge investments in Internet technology; .... , they have adopted e-mail, intranets, extranets, and customer-relationship management
    a) is to say   b) e.g.   c) in example   d) as regards

2. .... the ignition is switched on, a car cannot start.
    a) If   b) Provided   c) Otherwise   d) Unless

3.   .... or not the automated mechanism requires highly pollutant batteries is not important for the time being.
   a) Whether   b) Either   c) Both   d) Otherwise

4.   The project was sponsored by the Science and Technology agency. .... we would have had to abandon the project.
   a) Unless   b) If doesn't   c) However   d) Otherwise

5.   ...... the progress in this field, no device has been sufficiently developed for commercial application.
   a) Despite of   b) Although   c) Though   d) In spite of

6.   The inventions of .... the transistor in 1947 .... the integrated circuit in 1958 have made possible the development of small-size, low-power, low-weight, and high-speed electronic circuits which are used in the construction of satellite communication systems.
   a) not only/but as well   b) whether/ and   c) either / nor   d) both /and

7.   A plastic that is biodegradable has not yet been invented .... all the advances that have been made in the plastic industry.
   a) although   b) in spite of   c) owing of   d) since

8.   The mobile-Internet firms are carefully analysing this market ...... the mistakes made by fixed-line-Internet dotcoms are not repeated.
   a) in order for   b) so that   c) for that   d) so as not to

9.   The average molecular weight of paraffin is around 350 .... the average molecular weight of wax ranges from 600 to 800.
   a) however   b) whereas   c) nevertheless   d) otherwise

10.   Sensors will trigger the alarm .... there is no power failure.
   a) if not   b) even   c) as a result of d) so long as

11.   .... this is a highly flammable product, the company must warn the employees.
   a) Because of   b) Since   c) Due to   d) As a result of

12.   Someone tampered with the front door lock, ... activating the alarm.
   a) for   b) thus   c) consequently   d) as a result

13.   .... having struck a big iceberg, the oil tanker sank.
   a) Since   b) Therefore,   c) Due   d) As a result of

14.   .... electric cars may look complicated, they are in fact based on rather basic scientific principles.
   a) However   b) Despite of   c) Although   d) Whereas

15.   The latest Taft screens are not so harmful to human eyesight as traditional ones and at the same time are quite inexpensive, ... their popularity.
   a) since   b) owing to   c) because of   d) hence

16. Special prices at the faculty bar are offered to university students only, ... people between 18 and 24 of age.
    a) such as   b) e.g.   c) i.e.   d) otherwise

17. The transport of dangerous waste will be done in two ways, ... train or ship.
    a) namely   b) as   c) in summarising   d) in addition

18. .... the idea of treating different language learners differently is quite simple, implementing a one-to-one syllabus is not.
    a) While   b) However   c) Despite   d) In order to

19. Total demolition, ... demolition where the whole structure is removed, is a type of demolition used when clearing sites.
    a) that is to say   b) for example   c) such   d) moreover

20. The process of writing basically consists of three main stages, ..., the pre-writing stage, the writing stage, and the post-writing stage.
    a) as   b) namely   c) in summarising   d) in addition

21. At first sight, one expects that the mobile Internet is mostly used in person-to-person communication. The fancier possibilities of the mobile Internet are in fact adopted by businesses, ...
    a) although   b) despite   c) though   d) thus

22. Some people had warned me against the disappointing conclusions in the report; ...., I didn't expect them to be so bad!
    a) although   b) therefore   c) yet   d) in consequence

5-6 Read these incomplete sentences and then try to write up full sentences with the help of the following correlative structures: *both…and, either…or, neither…nor, not only… but also,* and *whether…or.* An example is provided below.

Example: Metals / can be / ferrous, non-ferrous: *Metals can either be ferrous or non-ferrous.*

1. This product / be / not flammable, radioactive.
2. Stability, malleability / be / the properties we require.
3. Students like it / students not like it / they must take an exam at the end of the semester.
4. In semi-virtual courses / home assignments / can be submitted in class / can be sent via the intranet. Students can choose the system they prefer.
5. It is argued that industrial injury rates / depend on economic factors / depend on the role of the State and safety engineers, too.
6. Unemployment / workers' reluctance to report accidents / not help to reduce workplace safety.

7. The managers interviewed / be involved in accident prevention / show high tolerance for rule violation.
8. Some foreign language learners feel that / learning words by heart / doing mechanical exercises / not provide them with enough fluency.
9. It remains to be seen / students coming from private secondary schools outperform students from state secondary schools at university.
10. Research on composites / be necessary in the fields of construction materials and architecture / be very necessary in the fields of biotechnology and medical surgery.

5-7 Read the following text and fill in the blanks with suitable connecting expressions. Note that you can't make any changes, can't use SO, BUT, AND, IF, and RELATIVE PRONOUNS or repeat connectors.

In the Pursuit of Energy

...... 1 ...... crude oil prices steadily rise above 40$ per barrel, the power equation seems to be shifting and viable alternatives seem to be emerging. It's not only Western economies that are thirstier than ever for energy, ... 2 ...... emerging economies ...... 3...... China and India, which are intensifying pressure on petroleum supplies and prices.

One of the first alternatives is nuclear power. Nuclear power was once shunned, ... 4 ...... it is now resurging as an increasingly popular option ...... 5 ...... a growth of domestic electricity; more specifically, in the US domestic electricity demand will grow by 20 percent in the next twenty years. ...... 6 ...... of this, nuclear power is considered attractive ...... 7 ...... we are not yet running out of uranium supplies, it is quite inexpensive and it does not emit greenhouse gases or cause any of the environmental damage related to fossil fuels. Therefore, the low operating cost and the small environmental footprint of the nuclear industry seem to be a winning combination.

Natural gas is and has been the fuel choice for this decade. Natural gas used to have several logistical problems associated with its transportation and its uncertain supply. ...... 8 ......, new cost-effective technologies that cool natural gas to a liquefied state have been developed ...... 9 ...... liquefied natural gas (LNG) can be more easily transported via specially-built ships. Japan, Taiwan and China are already large importers of LNG, and ...... 10 ...... terminals where tankers off-load the fuel may be getting small soon.

In addition to nuclear power and natural gas, there have also been positive trends in the coal industry. ...... 11 ...... from its low price, groundbreaking research on clean(er) coal technologies should be singled out. Coal supporters predict that the coal shares of the electricity market will increase ...... 12 ...... more, particularly if the tools to transform coal from the 20th century polluter into a 21st century environmentally-friendly energy source are eventually provided.

Next there is hydrogen, the earth's most abundant element. Hydrogen is expected to power future cars and to become a leader in this new playing field. ...... 13 .... oil giants like BP are involved in breakthrough research, ... 14 ...... corroborating this optimism about hydrogen, we should be cautious. Two main roadblocks still exist.

First, automobile fuel cells need to be more durable, reliable and cheaper and, ... 15 ......, the logistics of transporting hydrogen and making it available for mass-market use also needs a solution.

...... 16 ......, whether petroleum prices continue rocketing or not, oil is still the king. The demand for oil from industrialised Asia is going to double in the next few years and new joint venture deals with Asian's state-owned oil companies have already occurred. We have been told over and over again that *today's* oil reserves will be extinguished in about forty years. .... 17 ......, there is much more oil which has not yet been discovered or which is known to exist but in extremely deep waters and/or in extremely cold or hot climates. Technologies to obtain oil under these more complex conditions have not yet been developed. Nobody believes that engineers in the 21st century are less skilful than in the past, ... 18 ....

Adapted from: Energy: a new playing field by R. McGarvey, *Harvard Business Review*, May 2005. Reprinted with permission of the publisher.

5-8 Complete the sentences below with a suitable connecting expression.

1. .... their growing population, some Eastern countries need more food.

2. The interns in the department often do routine secretarial-type work; ...., they often prepare coffee and tea for their bosses.

3. The dean agrees with your suggestion for our faculty; ...., that smoking should be banned everywhere.

4. The function of a scientist is to know .... the function of an engineer is to do.

5. He wore special clothes in the laboratory .... disobey the orders.

6/7. Banks and other businesses .... restaurants, bars and other shops were making the necessary arrangements in 2002 .... the prediction that people would prefer credit cards to Euro currency came true.

8/9. .... many different types of gaskets or nuts and bolts are stocked, sooner or later the wrong type will be installed. It is much better to keep the number of types stocked to a minimum .... minimize errors.

10. Economies of scale can generate cost advantages for large firms; ...., important diseconomies of scale can actually increase costs if firms grow too large.

11. There are some important physical limitations to the size of some manufacturing processes;..................., engineers have found that cement kilns develop unstable aerodynamics above seven million barrels per year capacity.

12/13. Very specialized jobs can be very disappointing for employees. .... workers become mere 'cogs in a manufacturing machine' worker motivation wanes, ... affecting productivity and quality.

14. .... of competing with Nike, Reebok, and other high-priced shoe firm, Addidas has decided to sell its basic shoes at relatively lower prices in such a highly competitive market as the US market.

5-9 Read the following pairs of sentences and think how they could be joined by means of a connector so they make sense. You can make as many changes as necessary, but remember the aim is to practise connecting expressions.

1. We should send the feasibility report to our white-collar staff. Our white-collar staff is made up of female engineers aged between 25 and 35.

2. The protective ozone layer has been seriously damaged in the past decades due to the emission of chlorofluorocarbons. Humans should stop using them.

3. It is necessary to bear in mind the unethical uses of genetics. It is impossible to understand why some governments are reluctant to regulate the scientific use of genetics.

4. Some raw materials will exhaust in a few years. Governments must do something to prevent their extinction.

5. A computer has a far bigger memory than a human being. A computer has to be programmed by a human.

6. Benzene is the most feared of the specific pollutants emitted by the motor. Benzene has an incontestable toxicity.

7. E-mail is an Internet resource which provides people with maximum utility and ubiquity. E-mail requires a permanently updated security appliance to be protected against viruses.

8. Skylab was to be a laboratory in orbit where men could live and work for extended periods. The initial concepts of its functions and uses evolved and changed with the passing of time.

9. The willingness to pay a premium for renewable energy and energy efficiency among the general public can be explained by demographic factors. Among these factors we find age, salary, and education.

10. The truck drive-shaft spline failed. Corrosion fatigue caused this failure.

11. The piece has failed. One of its parts has become permanently distorted and its reliability has been downgraded.

12. A special wiper system has had to be designed. The wiper must clean water and debris from the angled windshield of this aerodynamic car.

13. We know exactly which torque has been used on this joint. We don't know the exact preload created by that torque.

14. Certain metals should be protected from the rain. They can get rusty.

15. The structure can break. The pressure must be reduced.

16. Hewlett Packard makes a wide variety of testing and measuring instruments, ranging from $400 oscilloscopes to $500 microchip-testing systems, which require regular maintenance and calibration. Hewlett Packard provides the maintenance and calibration service to its customers.

17. The Internet is a technology for interacting with customers. It is the interaction itself that creates a valuable relationship between a company and its customers.

5-10 Finish each of the following sentences in such a way that it means the same as the sentence printed before it.

1. The chassis will be easily corroded if you do not use a special paint to protect it.
   Unless...

2. The reason for the misunderstanding was the bad transcription of the recorded information.
   There was a misunderstanding *as…*

3. In spite of the enormous effort made, they still couldn't finish the project on time.
   Although...

4. I'll send you another copy of my book since it is probable that you have lost the one I gave you.
   In case...

5. As a consequence of the unexpected results, the laboratory is considering the possibility of repeating the test.
   Because…

6. He telephoned to find out what was wrong with the shipment.
   He telephoned *so…*

7. Governments are not taking the atmospheric reduction of pollutants seriously enough; consequently, global warming will keep on increasing.
   As...

8. The fuses blew several times due to a loose connection.
   Since…

9. The firm has improved the production method and revised the prices; however, the sales are still going down.
   In spite of …

10. We use a telescope to see distant bodies.
    A telescope is…

11. The lecture began at 7 o'clock and we arrived just after that.
    When…

12. They couldn't reach the other side because it was too far away.
    If …

13. Metals conduct electricity, plastics cannot.
    Unlike...

14. You will not get the desired printing quality unless you replace the ink cartridge.
    Provided...

15. First, the light turned red and then the alarm rang.
    After…

5-11 Read this poorly organized essay written by a student and improve it in two ways. First divide the text into the number of paragraphs you think necessary. Then provide the missing transition signals so that paragraphs are better tied up. Finally, choose the best location for the transition paragraph given below.

Atmospheric pollution

Air is ninety percent nitrogen, oxygen, water vapour and inert gases and we breathe it because it gives us the oxygen that is essential to live. Nowadays, however, air is composed of other substances that are released into the air by human beings and which cause air pollution. There are several consequences derived from air pollution which have serious implications for our health and well-being as well as for the environment. Before analyzing these consequences, however, it becomes necessary to underpin the main causes of air pollution. The release of tiny particles, measuring about 2.5 microns or about .0001 inches and produced by the burning of fuel in automobiles, homes and industries, becomes a major source of atmospheric pollution. Besides, other noxious gases, e.g. sulphur dioxide, carbon monoxide or nitrogen oxide that take part in further chemical processes and reactions, are also expelled to the atmosphere. Some chemical reactions between pollutants derived from different sources (for example, automobile exhaust and industrial emissions) cause an effect known as smog. Depending on the geographical location, temperature, and wind and weather factors, pollution is dispersed differently. In other words, when the air close to the ground is cooler than the air above, temperature inversion takes place. Under these conditions, polluted air cannot rise or disperse and, as a result, atmospheric pollution can increase up to dangerous levels. Winter inversions are likely to cause particulate and carbon monoxide pollution; in contrast, summer inversions create smog. For this reason, many cities that are centres of pollution suffer from the effects of smog, especially during the warm months of the year. Another consequence of air pollution is acid rain, which is produced when a pollutant like sulphuric acid combines with drops of water in the air and acidifies these water drops. The effects of acid rain on the environment can be very harmful because they damage plants and poison the soil, changing the chemistry of lakes and streams. Among other things, acid rain can also kill trees, animals, fish and, in a word, any type of wildlife as we know it today. The greenhouse effect is another result of air pollution, also referred to as global warming. This effect is related to the building up of carbon monoxide gas in the atmosphere. Carbon dioxide is produced when fuels are burned. Plants in fact convert carbon dioxide back to oxygen, but the release of carbon dioxide caused by human activity is higher than what the world's plants can process. The situation is worsened since many of the earth's forests are disappearing, thus making the amount of carbon dioxide dramatically increase. This effect causes terrible changes in the climate: if global temperatures are higher, the polar ice may melt and many cities would remain under sea level. Another effect is the ozone layer depletion. Chemicals are expelled to the stratosphere, one of the atmospheric layers surrounding the earth; as a consequence, the ozone layer in the stratosphere, which protects the earth against harmful ultraviolet radiation from the sun, is damaged. Chlorofluorocarbons (CFCs) from aerosol cans and cooling systems reach the stratosphere and cause holes in the ozone layer, thus allowing radiation to reach the earth. Ultraviolet radiation is known to cause *skin cancer, to have damaging effects on plants and wildlife, and to collaborate to global warming. Nowadays many governments of the world are taking steps to stop the damage from air pollution. However, we individual citizens must cooperate in this arduous task. We should try to make everybody aware of this problem and do everything in our hands to minimize the effects of air pollution. If everyone does one's bit, we will manage to somehow combat pollution and save our planet.*

Transition paragraph:

As we have seen in the paragraphs above, atmospheric pollution can affect seriously our health in different ways and, ultimately, our survival as a human race. Smog, acid rain, global warming, and the widening of the ozone hole are so pervasive that a complete eradication of atmospheric pollution is practically impossible; yet we can reduce and prevent it from growing. A thorny question arises at this stage: what actions can be taken to reduce and prevent air pollution? This is a complex and interesting question which we are not going to address here.

5-12 Carefully study this outline where the introduction, the topic sentences and the conclusion have already been written. Write both short and long transition signals so that the ideas in the body paragraphs are tied up and cohere.

Industrial Design

Everything that man makes is designed, but not everything is well-designed. Good design only comes about when products are made with attention both to their functional and their aesthetic qualities. Objects must do the job they were intended to do, and they must be well-made and efficient; yet they must also be pleasing to use and to the eye. This is probably what Edison meant when he said that good design is ninety percent commonsense and two percent aesthetics. This excerpt is largely about the design of the objects we have in our home and workplace and the vehicles which carry us around, but first let us have a look at how the standards and expectations about design have been constantly evolving.

I. The term design is itself a modern invention, a product of the British industrial revolution, which began in the late 18th century.
    Supporting sentences : industrial process of breaking down manufacturing into its component tasks.

II. Design is a product of the British industrial revolution rooted in the introduction of novel techniques of high-volume production.
    Supporting sentences: correlation between high volume and low unit costs. Consequence: long lead times and standardized products.

III.   Engineers and entrepreneurs in Britain were the first to notice and to exploit the technological and social changes that arose from this machine age revolution.
Supporting sentences: a careful product planning was necessary → the person who created an object in his/her mind was no longer the same person who built or manufactured it.

IV.   The great achievements in design of the 19th century were mainly big buildings and industrial facilities.
Supporting sentences: Some examples are: civic monuments, bridges, structures and steamers of Telford, Paxton and Brunel.

V.   The achievements of the 20th century are far more personal, even domestic, in scale.
Supporting sentences: Some examples of Modernism and Functionalism are: Ferdinand Porsche's Volkswagen, or the Sony 'Walkman'.

VI.   In today's mass market, every consumer can be a design critic and the more discriminating people become, the more manufacturers realize that merchandise must meet the demands made of it.
Supporting sentences: Post Modernism. The language of objects: not only scientific considerations but also emotional, symbolic and metaphoric. → consumer psychology in product design and development and marketing.

During these one hundred and fifty years, mass-production has evolved and has been perfected, and in the course of this evolution the designer has been variously an artist, an architect, a social reformer, a management consultant and an engineer. Design will change in accordance with technology and social conditions but in important respects it will always remain the same. This is so because the designs of the past have contributed ideas about form which are a part of our current language of objects and which are the bases for future design solutions. New developments in technology are going to mean an enhanced role for design engineers.

5-13 Read this essay written by a student and answer the questions below.

a) Circle the short and long transition signals you can identify.
b) Identify the transition sentence that goes over one paragraph.
c) Identify the use of key words, pronouns or synonyms that enhance inter-coherence.

Pipelines made of PVC

One of the most important, widely used and oldest plastic is PVC (Polychloride of Vinyl). PVC is a polymer which contains approximately 56% of chlorine that only with the use of certain additives could be transformed in a material suitable for different uses. Several additives allow a large variation in PVC characteristics and a high correspondence with the required application. Owing to these characteristics, there are two main common sorts of PVC that could be used for fluid pipelines: PVC-U and PVC-C.

PVC-U and PVC-C are two subtypes of PVC that share some similarities but they have some distinctive features that explain why each is appropriate for different uses. To begin with, PVC-U is a polyvinylchloride, the plastic most often used in fluids channelling since it is rigid and versatile. Its application procedure fulfils the following requirements depending on its resistance or installation. As to resistance, PVC-U avoids graze, corrosion, chemical inorganic substances, acid and alkaline solutions, and aromatic chlorinated hydrocarbons. Regarding its method of installation, PVC-U is ideal for exterior facilities because it withstands temperatures ranging between 32°F and 140°F and no special tools are needed for the joints given that these joints are adhesive and elastic.

On the other hand, PVC-C is a good substitute for conductions in the chemical industry or other industrial applications. PVC-C is realized by means of post-chlorination of the PVC. PVC-C is a transformation of the PVC-U and, due to its chemical formulation it can withstand higher temperatures than the PVC-U (ranging from 32°F and 176° F), with the consistent increase of tension force, good tenacity and excellent chemical and mechanical resistance. As with PVC-U, it is not necessary to use special tools for PVC-C installation and it is resistant to a lot of chemical inorganic substances. In addition, it has a low thermal conduction, it is not flammable, or resistant to organic solvents, and it is auto-extinguishable.

As a result of all the characteristics described above, PVC is the ideal material to satisfy pipeline requirements. In brief, the reasons why PVC is used for fluid pipelines are that it complies with the temperature work rank, it has a long life cycle (it is neither practical nor economical nor ecological to replace it), it resists chemical agents' attacks and it is inert. In conclusion, PVC is the material that best suits pipelines' requirements.

5.4 Revision of grammar and style

In this section you will find some further exercises to improve your command of grammar and to practise different aspects related to style. If you want to revise the corresponding theoretical explanations you can refer back to sections 4.3 and 4.4 in Chapter 4.

Grammar

INDIVIDUAL TASKS

5-14 Read the following sentences, looking for a grammatical mistake of any type and then provide a correct version without altering the meaning. Every sentence contains only *one* mistake.

1. The two maintenance heads enjoy meeting and playing cards at the week ends. Who do not get on very well at work.
2. Every Christmas card is always signed by the two managers, who considers this time-consuming activity is worthwhile and necessary.
3. The firm negotiated on a price and delivery terms basis. Because the firm wanted the most reliable supplier.
4. On the leaflet, it encourages university students to take as many English courses as possible.
5. Everyone is entitled to their pension scheme in this company.
6. The battle of the quality department are frequent complaints.
7. Written in red letters, they read the words 'No mobile phones'.
8. The driving force of these polymerizations with a sequential repeat unit correspond to the formation of metal halide salts.
9. While he was writing the conclusions. His friend was preparing the PowerPoint slides for the oral presentation they were supposed to give.
10. Either of these cellular phones work well so long as you handle it with care.
11. The three packets have to be shipped back to its manufacturer as soon as possible.
12. If the rod glued to the casing were exposed to excessive heat, it would impair the security of the structure.
13. The company have organized a celebration to honour its three oldest employees.
14. Some people thinks that his career is the only important thing in life.
15. Neither his father nor his brothers is a member of the society.
16. Remove the fine-grained sand from the liquid and analyze it.
17. The company launched a new product into the market which it was expected to make the company win the market share.
18. Flying in the sky, I saw a hydroplane.
19. Does people always behave rationally and coherently?

20. Graham recommended a total artificial hip replacement for the injured child, a highly reputed specialist in biocompatible implant materials.

21. The item was shipped to the client, without noticing it was defective.

5-15 Read the following passages and proof-read them for any grammatical mistake. For the purpose of conciseness, also reduce the relative clauses in the passages whenever possible.

Passage 1

It is apparently difficult to estimate the cost of an R&D project in the screening process (1). In addition to the uncertainty of estimation, is thought that inaccuracies also stem from deliberate under-estimations used to marshal support for a project (2). It was found that the ratio of actual to estimated costs were greater for project failures than for successes (3). One explanation may be the tendency not to give up a project and continuing to allocate funds to this project even when success seems unlikely (4). Although costs are difficult to estimate; probabilities are even more difficult (5). Similarly, the project originators and the persons with implementation responsibility tends to give more optimistic estimates than the average whereas those with 'a knowledge gap' about technical feasibility tend to give more pessimistic estimates (6). Therefore, estimates should be treated with extreme caution and you should take steps to minimize organizational bias on the costs and probabilities (7).

Adapted from: Urban, G & J. Hauser. (1993). *Design and marketing of new products*. New Jersey: Prentice-Hall,158-9. Reprinted with permission of the publisher.

Passage 2

A quick look under the sink or in the garage of your home can reveal the number of products that exist to make tasks such as cleaning and polishing easier (1). During the first half of the twentieth century, Americans did not have the vast selection of consumer products which contain chemicals that we now do (2). Think about the numerous items that perform useful services for us; such as oven cleaners, toilet cleaners, floor strippers, and the list goes on and on (3).

What is surprising are that many of the materials we work with or use daily are hazardous (4). As long as we use them properly, under normal circumstances, they should not pose health problems or environmental damage (5). What makes a material hazardous is their basic chemical, physical or biological characteristics (6). If a material catches on fire easily or spontaneously under standard temperature and pressure conditions is flammable (7). If a substance reacts violently by catching on fire, exploding, or giving off fumes when it is exposed to water, air, or low heat, it is reactive (8). If a material releases pressure, gas, and heat suddenly when subjected to shock, heat, or high pressure, it is explosive (9). Lastly, if a material emits harmful radiation, it is radioactive (10).

Many of the chemicals that exhibit these hazardous characteristics are not readily biodegradable (11). This means that they are not able to be broken down easily into their component parts by micro-organisms (12). This types of chemicals are said to be persistent in the environment because, once released they tend to stay intact and

> possess the same dangerous properties for long periods of time (13). Toxic substances that occur naturally can also pose problems (14). A wildlife refuge in California became contaminated by selenium, an element commonly found in high pH desert soil (15). The build-up of the selenium occurred because of the irrigation methods which were used by nearby farms (16). During irrigation water carried dissolved selenium to the wildlife refuge (17). Deformities in developing embryos were found more frequently that had ingested selenium (18).

Adapted from: Boyce, A. (1997). *An introduction to environmental technology.* New York: Van Nostrand Reinhold, 265-266.

5-16 Read the following sentences and identify the type of mistake in each of them. Then choose the alternative that you consider most grammatically correct and that doesn't alter the meaning of the original sentence. If you believe there is no error in the original sentence, then you should choose option *a*:

1) Having missed many class tests, *it was impossible to pass the final examination.*
   a. it was impossible to pass the final examination
   b. she could not pass the final examination
   c. the final examination was not passed
   d. passing the final examination was impossible

2) *Sensitive to his co-workers' motivation and formation, an in-company formation plan was offered to trainee engineers.*
   a. Sensitive to his co-workers' motivation and formation, an in-company formation plan was offered to trainee engineers.
   b. Sensitiveness to his co-workers' motivation and formation; an in-company formation plan was offered to trainee engineers.
   c. Sensitive to his co-workers' motivation and formation, Mr Hewings offered an in-company formation plan to trainee engineers.
   d. Because of he was sensitive to his co-workers' motivation and formation, an in-company formation plan was offered to trainee engineers.

3) Each of the two material strengthening methods *which they were proposed were finally rejected.*
   a. Each of the two material strengthening methods which they were proposed were finally rejected.
   b. Each of the two material strengthening methods which they were proposed was finally rejected.
   c. Each of the two material strengthening methods which were proposed was finally rejected.
   d. Each of the two material strengthening methods proposed were finally rejected.

4) The incidence of responsibility and salary *correlates positively with the degree of people's perception of work satisfaction.*
 a. The incidence of responsibility and salary correlates positively with the degree of people's perception of work satisfaction.
 b. The incidence of responsibility and salary correlate positively with the degree of people's perception of work satisfaction
 c. The incidence of responsibility and salary correlate positively; with the degree of people's perception of work satisfaction.
 d. The incidence of responsibility and salary correlates positively; with the degree of people's perception of work satisfaction.

5) Having examined the reasons for student drop-outs, *they concluded that they neither have the authority and that they do not have the means to reduce drop-outs, either.*
 a. Having examined the reasons for student drop-outs, they concluded that they neither have the authority and that they do not have the means to reduce drop-outs, either.
 b. Having examined the reasons for student drop-outs, it was concluded that they neither have the authority and that they do not have the means to reduce drop-outs, either.
 c. Having examined the reasons for student drop-outs, it concluded that they neither have the authority nor the means to reduce drop-outs.
 d. Having examined the reasons for student drop-outs, they concluded that they neither have the authority nor the means to reduce drop-outs.

6) The Executive Board of big companies like 'WriterCo' *may decide which of their managers in sister companies should be invited to resign.*
 a. The Executive Board of big companies like 'WriterCo' may decide which of their managers in sister companies should be invited to resign.
 b. The Executive Board of big companies like 'WriterCo' may decide which of their managers in sister companies invited to resign.
 c. The Executive Board of big companies like 'WriterCo', it may decide which of their managers in sister companies should be invited to resign.
 d. The Executive Board of big companies like 'WriterCo' deciding which of its managers in sister companies should be invited to resign.

7) *In the figure above, it is shown the process of the production of paper.*
 a. In the figure above, it is shown the process of the production of paper.
 b. It is shown in the figure above the process of the production of paper.
 c. The process involved in the production of paper it is shown in the figure above.
 d. In the figure above, the process of the production of paper is shown.

8) *The electrical installation is not supplying enough power to the tourist zone, it should be changed before summer time.*
   a. The electrical installation is not supplying enough power to the tourist zone, it should be changed before summer time.
   b. Unable to supply enough power to the tourist zone, we should change the electrical installation before summer time.
   c. Since the electrical installation is not supplying enough power to the tourist zone, it should be changed before summer time.
   d. The electrical installation is not supplying enough power to the tourist zone so they should be changed before summer time.

9) Each of the ten laptops he bought *was fast and ergonomic.*
   a. Each of the ten laptops he bought was fast and ergonomic.
   b. Each of the ten laptops he bought were fast and ergonomic.
   c. Each of the ten laptops which it was bought was fast and ergonomic.
   d. Each of the ten laptops which bought was fast and ergonomic.

10) *Provided cables with connections between telephone exchanges* tend to be coaxial.
   a. Provided cables with connections between telephone exchanges tend to be coaxial.
   b. Provided with connections between telephone exchanges, they tend to be coaxial.
   c. Providing connections between telephone exchanges, this tends to be coaxial.
   d. Cables providing connections between telephone exchanges tend to be coaxial.

Style

INDIVIDUAL TASKS

5-17 Read the following overloaded sentences and rewrite them into shorter and clearer ones.

> Smog does not only damage crops but also damages forests and pasture lands that produce another $700million in revenue for California each year, and that account for approximately 85 per cent of the California's land area and provide recreation and watershed land as well as support for the timber and livestock industries.

> Adapted from: Boyce, A. (1997). *An introduction to environmental technology.* New York: Van Nostrand Reinhold, 150.

> The fundamental building block of all organisms is the cell which is composed of 85 to 90 per cent of water and a variety of chemical elements, carbon being the most important of them and carbon combines with itself to form long chains of carbon atoms and combines with other elements to form the various organic molecules typical of living organisms.

> Adapted from: Boyce, A. (1997). *An introduction to environmental technology.* New York: Van Nostrand Reinhold, 108.

In comparison with older tanks made of bare steel that failed primarily due to corrosion, new tanks, which use new materials that began to appear in the 70s to minimize failure, and that are coated and protected from corrosion or are made of new non-corrosive materials such as fibreglass reinforced plastic, have nearly eliminated failures induced by external corrosion.

Of the existing fresh water sources only some of them are usable and available in limited quantities and are considered precious resources that are increasingly threatened by the effects of human activities such as the application of pesticides and fertilizers to the land, uncontrolled hazardous waste disposal, leaking underground storage tanks, and the use of septic tanks and drainage wells.

Adapted from: Boyce, A. (1997). *An introduction to environmental technology*. New York: Van Nostrand Reinhold, 159.

5-18 According to what you have learned about the use of the passive form decide which of the following sentences should be changed, either from active to passive form or vice versa. Remember to justify the change.

1) Somebody decided to postpone the meeting to the following week.
2) Thousands of tones of used paper are thrown away every day.
3) Many silly mistakes are being made by learners because of their inexperience.
4) The librarian hasn't organized the books in the library yet.
5) The modules themselves were initially developed without any definite physical characteristics.
6) They needed expert mechanics to adjust the compression mechanism.
7) To meet a more specialized need, the DS1846 chip was developed by Dallas.
8) The author of the paper devoted half of it to explaining the method used.
9) A practical example can illustrate the advantage of a high speed microcontroller.
10) Owing to her excellent work, Sue's boss promoted her to a higher position.

5-19 Read the following text and identify the informal and inaccurate expressions used. Now replace these expressions by more formal or technical ones. Some of them have already been marked for you.

The letters "FRP" are given to any number of advanced composites materials that use up high strength, oriented fibers that are either woven or twisted together or simply placed side by side in a lower strength polymeric matrix. In the case of most civil engineering applications, FRP usually *touches on* a class of stuff that uses E-glass fibers, in either a woven matt, roving, stitched piece of cloth, or a combination configuration. The glass fibers are frequently *soaked in* an isopoliester based-thermo-set matrix. As a result of this combination of composite material elements, the *technique of production* most frequently used up in advanced civil engineering composites is pultrusion.

> *Let's take a brief look at* the pultrusion process. In this process, spools of roving, strand, and pieces of cloth are pulled through a *widget* that wets and *soaks in* the fibers with polyester resign so that when the fibers are oriented and after this pulled thorough a heated die, the thermo-set matrix is *made up*, compressed and *made harder* as a result of chemical reactions that take place in and around the fiber. The end result is a shaped FRP component that can be made to any length since what comes out of the heated die is a continuous stream of FRP material. A large saw is typically used to cut the FRP elements into discrete member lengths as *said* by a given design application.

Adapted from: Keelor, D. C. et al. (2004). Service load effective compression flange width in fiber reinforced polymer deck systems acting compositely with steel stringers. *Journal of composites for construction.* ASCE, July/August, 290. With permission from ASCE.

5-20 Revise the sentences below to eliminate needless prepositional phrases, nominalizations, redundancies and wordy phrases.

1. In reference to the profit sharing plan the new manager has reached the decision, in opposition to the workers' opinion, that they should be reinvested in more modern machinery.
2. Unless we apply a particle accelerator and higher temperature to the experiment, this is bound to end up with a failure.
3. Please carry out a survey and make a report of the new situation.
4. In the event that the pressure is kept too high for a long period of time, the system runs the risk of undergoing an explosion.
5. The use of the safety belt together with the air bag system unavoidably leads to a reduction of injuries in car accidents.
6. As far as writing is concerned, computers are associated, in a favourable way, with producing texts in a faster and clearer way.
7. My interest in satellites was due to my discovery of the applications they have in our daily life.
8. Despite the proper working of the system at the present time, it has to be revised periodically in order to avoid that more accidents may happen in the future.
9. There is the possibility that we incorporate a considerable amount of workforce into our branch in Poland.
10. It is obvious that the firm's complaint about the price that was agreed upon and that has not yet been changed has become a main concern to the director.

5-21 Below there are some sentences taken from technical descriptions written by students. They contain different types of mistakes such as an inaccurate use of vocabulary or grammar and an incoherent use of style. Revise them and write an improved version.

1. Always make sure that car tyres are in good pressure.
2. … a nice, white-painted French vehicle.

3. But, anyway, this car is…
4. All technical progresses are installed in the car.
5. The car is small and it is very easy to fill the boot.
6. The documents can't be kept in the glove compartment and I have to put them in a box.
7. I fastened the seatbelt because it has been demonstrated that this can save my life.
8. The bumper is there for crashes that might come.
9. Safety systems have been developed in the interior and exterior of cars.
10. The car is known to be very short in terms of height.
11. The people inside the vehicle….
12. The steering-wheel is one of the most important things of a car.
13. The direction is given by the steering-wheel with the movement of the hands.
14. Passengers are fixed by means of the seatbelts.
15. Driving without the seatbelt is believed to be a risky quality.
16. Internet has become one of the most important information technologies. That's why its possibilities are really enormous.
17. Nowadays we can enjoy the capability to be able to be connected to the whole world with our computer.

5.5 Punctuation

Although some people disregard punctuation because they think it is a minor aspect of writing, punctuation is as important as intonation is in oral language since it contributes to making the writer's meaning clear and eases the reading. It is much more than a set of fixed rules and much more than a few marks used to divide sentences. Punctuation acts as the regulator of the flow of information through a sentence, thus helping readers to better understand the writer's ideas. Bear in mind that while there is a certain amount of subjectivity in punctuation, there are some rules that should be kept if we want to make the reader's task easier. In short, we can say that punctuation marks should be used according to fairly strict conventions.

Period (.)

The main function of the period is to indicate that a sentence has reached its end. British English uses the expression "full stop" to clearly indicate so. The period is also used in abbreviations, decimals and amounts of money. Periods are mainly used:

- to mark the end of a declarative and imperative sentence:

    e.g. The oscillator and the amplifier are combined in a single chip integrated circuit.

e.g. Make sure that you don't go into the red or you'll get distortion.

- to indicate abbreviation:

e.g. dept.   incl.   km.   cm.   p.   ref.

Some abbreviations that come from a two or three-word expression are written with internal periods:

e.g. a.m.   p.p.   i.e.   Ph.D.   p.t.o.   g.p.o.

However, names of organizations (NATO), government agencies (CIA), acronyms (LASER), Internet abbreviations (URL), abbreviations for states (NM) or time indicators (BC) are written without any period.

- to indicate decimals and amounts of money over a dollar:

e.g. $8.9 $6.75 $ 100.25

- to introduce quotation marks. Note that the period is placed inside the quotation in Am. English but outside the quotation in Br. English.

e.g. Tom said," I have almost finished the project." (Am. English)

Tom said, "I have almost finished the project". (Br. English)

Comma (,)

A comma, as opposed to a period, separates parts of a sentence, not sentences from each other. It is the most widely used punctuation mark and also the most flexible; hence its difficulty in using it. Commas are mostly used:

- to connect independent sentences joined by a conjunction (*and, but, for, or, nor,* so, *yet*):

e.g. The metal detector was developed for military purposes, but nowadays it is also used to locate pipes, cables and lost valuables.

Too little water leads to complete bonding, and too much results in excessive porosity.

When the sentences are short and joined by *and* then the comma is omitted. But if the sentences are joined by *but* or *and yet*, the comma must be kept:

e.g.   The room was dark and the speaker could not give the speech.

The room was dark, and yet the speaker gave the speech.

The FBI could scan the tag, check it and track it down to the store.

The comma is also omitted if the independent sentences have the same subject (or compound predicate):

e.g.   The GPS was developed by the US military and it was designed to pinpoint locations.

- to set off most introductory dependent clauses:

e.g.   When ground waves pass over sand, they lose energy.

If plasma technology became commonplace, it would result in significant implications for electricity generation and distribution.

- to set off adverbial phrases at the beginning of the sentence:

e.g.   After feeding the rays into the microchip, they are interpreted and verified.

Due to an estimated growth of domestic electricity, nuclear power is now resurging as an increasingly popular option.

If the introductory adverbial phrase is short, then the comma is omitted:

e.g.   In a future a phone book will no longer be a paper book.

For the last two decades digital technology has slowly invaded television.

- to set off parenthetic (non-restrictive) information that has a close relationship with the rest of the sentence. Parenthetical information that has a more remote logical relationship with the rest of the sentence should be set off by dashes or parentheses:

e.g.   Noise, which is any unwanted signals, can be a problem with amplifiers.
My colleague, John Herbert, recently won an important prize.

Digital and flat models—precisely the category targeted by the emerging nanotube technology—are the fastest-growing category.

- to separate coordinated items in a series, whether the items are words, phrases or clauses:

e.g.   There are three main types of transmission line cables: parallel wires, twisted pair, and coaxial cable.

They include making decisions, setting priorities, working in teams, running meetings and negotiating.

- to separate two or more adjectives if each modifies the noun alone:

e.g.   The result was a dark, hazy pattern consisting of numerous wavy, concentric circles.

- to set off interjections, transitional adverbs and similar elements that represent a distinct break in the continuity of the thought:

  e.g.   Well, a program is just an algorithm but written down in a language that the computer can understand.

  Indeed, RF-Id tags are only one example of a coming wave of wireless communication technologies.

  On the contrary, they often concentrate on the aesthetic design of the physical kiosk.

  Samsung, for example, has already demonstrated a full-colour 38 inch field-emission display.

  Today, however, the capacity of 650 MB of storage is too limited for computer applications.

With expressions such as *that is*, *namely*, *i.e.*, and *e.g.*, the use of the comma will be subject to the magnitude of the break in continuity. If the break is great, then a semicolon, a dash, or a parenthesis may be used:

  e.g.   Some of the more common steels are classified according to carbon concentration, namely, into low-medium, and high-carbon types.

  The most spectacular of RF-ID tags will be large-scale systems that piggyback on cellular networks to locate any cell phone on the network; namely, in the near future, your whereabouts will not be a secret if you are carrying your cell phone.

- to set off quotations, maxims, or similar expressions from the rest of the sentence:

  e.g.   Almost in passing, he mentioned," It is a striking property of these diagrams that they constitute records of three-dimensional as well as plain objects."

  He ended up his talk by saying," If it isn't broke, don't fix it". (Maxim)

- to set off a question from the clause that introduces it:

  e.g.   I ask you, is this a creature equipped for the onslaught of the information age?

  He asked, can you be an engineer and not be able to write, speak or listen effectively?

Semicolon (;)

The semicolon is used to mark a greater break than that marked by a comma. It is used:

- to connect independent or coordinate clauses that have a closer link than that provided by a coordinate conjunction:

    e.g.   Then with the tail raised, the take-off begins; its purpose is to accelerate the airplane to a velocity at which climbing is possible.

- to connect independent clauses joined by a conjunctive adverb:

    e.g.   Today's oil reserves will be extinguished in about forty years; nevertheless, there is much oil which has yet to be discovered.

    In order to take proper advantage of the economics of integrated circuits, designers have had to overcome some serious device limitations; therefore, an understanding of device characteristics is essential in designing good integrated circuits.

    However, clauses introduced with *so* and *yet* can also be written with a preceding comma:

    e.g.   Nuclear power was once shunned, yet it is now resurging as an increasingly popular option.

    Sheetco has decided to acquire a new installation for the two product lines of the company and improve the customer service department, so it has become necessary to reduce the company's distribution channels with all these changes.

- to separate items in a series which has internal punctuation:

    e.g.   Other applications include DVD-audio, high-capacity, high-quality audio disk; VD-R, a write-once DVD format for high-capacity data storage in computing applications; and DVD-RAM, a multiple read-write format also used for high-capacity storage in computing applications.

- to separate clauses of a long compound sentence or clauses that are themselves subdivided by commas, even when a conjunction is used:

    e.g.   In the case of tagged razor blades, it is believed that the loss will be small and incremental; whereas with cellphones tracking, it could be substantial and potentially intrusive.

- to set off expressions such as *that is*, *namely*, *i.e.*, or *e.g.* if the break is greater than that signalled by a comma:

e.g.   Sometimes there are certain cases where the selection is predetermined by general conditions; e.g. a small oil engine, for which the fuel economy is not so important as simplicity, should be built on the two-stroke cycle with crankcase scavenging.

- to separate a quotation mark or a parenthesis from the following sentence:

e.g.   Professor Harris assumed that the students had read "The rise of the network society"; he referred to this book several times during the class.

Director Higgins had thought that the board would accept his decision (some members had assured him privately that they favoured such a decision); but it had to be postponed because of the union's disapproval.

Colon (:)

The colon is used to signal a greater separation than the comma and the semicolon but less than that indicated by a period. It also signals anticipation, that is, it introduces information that expands the preceding independent clause. Colons are used:

- to introduce explanation, expansion, or elaboration after an independent clause:

e.g.   The first feature is that the links in the net are preferentially attached: A router that has many links to it is likely to attract still more links.

The point remains the same: pay attention and respect the speaker.

Nowadays these clauses are separated more frequently by a semicolon than by a colon or are treated as separate sentences.
Notice that the part following the colon usually begins with a capital letter in Am. English, and particularly if it consists of more than one sentence.

- to emphasize a sequence in thought between two clauses that form a single sentence (or to separate one clause from a second clause that contains an illustration or amplification of the first):

e.g.   The students could have inhaled the toxic substance: this may be the reason for them being poisoned.

- to introduce a rule or principle, a formal statement, a quotation or a speech in a dialogue:

e.g.   Here is how the scheme works: When a suspect token arrives, the host system creates a very big number, called challenge, entirely at random and sends it to the token.

e.g.   Interviewer: What exactly do you mean by that?
Michael: I mean that it should be seen as an "enabling technology".

■   to introduce a list or a series:

e.g.   Speakers of hi-fi systems usually contain up to three individual units: a tweeter, a squawker and a woofer.

Two basic processes are used to manufacture continuous glass filaments: marble melt and direct melt.

However, if the list is a complement or object of an element in the introductory statement, then the colon is not used.

■   To introduce the enumerated items that come directly after the expressions *as follows* and *the following:*

e.g.   The desirable features of the duplexer are as follows:
(1) Cross modulation distortion is small and stop band attenuation is large.
(2) Pass band attenuation is flat and insertion loss is small.
(3) Spurious response is small.

Dash (—)

The dash is longer than a hyphen and even though there are different kinds of dashes which differ in length and in use, the most common one is the em dash (—). Dashes are mostly used:

■   to denote a sudden break in thought that represents an abrupt change in sentence structure.

e.g.   The emission of exhaust gases into the atmosphere—let me remind you there were no catalysts at that time yet—was partly responsible for the greenhouse effect in Los Angeles.

■   to add or insert an explanation, a parenthetical or a digressive element. The dash can sometimes precede such expressions as *namely, that is, i.e.* or *e.g.* (A comma could also be used in these contexts if the break in continuity was not so great).

e.g.   Wear corrosion deteriorated the T-beam—a beam which could no longer withstand that pressure.

Ceramics are commonly used in heat exchangers—when the material is going to be exposed to high temperatures—while carbon is preferable in bearings and gears—products that are exposed to friction and wear.

- to enumerate or define some element. (Commas and parentheses could also be used and, if at the end of the sentence, so could a colon).

  e.g.   Three main courses are mandatory for electronics engineering students—circuit theory, electromagnetism, and digital systems.

  A router—a piece of equipment which takes packages of data and sends them to where they are trying to go—is necessary to stop hackers.

- to introduce or separate a summarizing clause from a list

  e.g.   Bolts, nuts, fastenings and rivets—these are the topics the chapter deals with.

  An accountant knows very well that there are two main types of costs—direct costs and indirect costs.

Finally, it is not advisable to use more than one dash or a pair of dashes in the same sentence to avoid confusion.

Parentheses ( )

Like dashes and commas, parentheses are used to set off explanatory, amplifying and digressive elements, but with some slight difference. When these parenthetical elements hold some close logical relationship with the rest of the sentence, commas should be used; when the logical relation is more remote, then dashes or parentheses are preferable. Parentheses are then used:

- to introduce some parenthetical element.

  e.g.   A disadvantage of Fourier analysis is that frequency information can only be extracted for the complete duration of a signal (see Chapter Four).

  The effect of ultra-violet light on particles emitted by cars (carbon monoxide, nitrogen oxides and hydrocarbons) triggers off a number of complex chain reactions.

- to enclose parenthetical elements prefaced by *namely*, *i.e.*, *e.g.*, *that is*, provided the break in continuity is greater than that signalled by a comma.

  e.g.   Composite materials (e.g. fibreglass, concrete, plastics or ceramics) have played a major role in materials development for the last few years.

- to distinguish two overlapping parenthetical elements in combination with dashes, when both dashes and parentheses break sentence continuity considerably.

e.g.   Dr Higgins' latest work on acoustics—which was published by the (university) academic publisher—has been very positively reviewed by specialists in the field.

Among composite materials, fibreglass (which was first developed in the late 1940s) can be considered the first modern composite—in fact, it is still the most common.

- to enclose numerals or letters in a listing format.

e.g.   Glass fibers are unique materials that exhibit the properties of: (1) hardness, (2) transparency, (3) resistance to chemical attack, (4) stability, and (5) inertness.

When using a closing parenthesis, you should not use commas, semicolons or dashes. When using an opening parenthesis, you should not use commas, semicolons or dashes either, except if the parentheses are used to enumerate and list.

Brackets [ ]

Also known as square brackets, they are less common than parentheses and are mainly used:

- to enclose editorial interpolations, corrections, explanations, translations, and commentary in quoted material.

e.g.   Edison [sic] is said to have taken a tough stance.

The detection of aircraft *was first accomplished* in the USA *in the 1930*. [Italics added].

- to enclose parentheses within parentheses.

e.g.   A router (a piece of equipment which takes packets [packages of data] and sends them to where they are trying to go) can stop hackers, among other things.

- to enclose a phonetic transcription.

e.g.   An additional water reservoir [rezəvwa:] was built on top of the hill.

- to enclose phrases like *Continued from...*, or *To be continued*, which usually appear in italics and smaller type.

e.g.   *[Continued from page 32]*.

Hyphen (-)

The hyphen, which as you have noticed is much shorter than the dash, is used:

- to separate numbers that are not inclusive, like a zone code in telephone numbers, and to write numbers.

    e.g.   34-93-401-00-01

    thirty-four

- to divide words in noun compounds.

    e.g.   a two-pinned plug (compound noun formed by a number and a noun)

    an air-cooling device (a compound adjective ending with the present participle)

    a better-prepared engineer, well-known techniques, ill-informed study (a compound adjective beginning with adverbs like *best*, *better*, *ill*, *lower*, *little* or *well*)

- to join coequal nouns (unless the first noun modifies the second)

    e.g.   a scholar-athlete, an author-chef

Quotation marks (" ")

Double quotation marks are used:

- to quote phrases or expressions from another context.

    e.g.   XYZ company offers a "legendary" rapid turnaround on high-speed machine aluminium tooling.

    An Mt can "switch on the fly" between different data channels without losing frame synchronization.

- to quote words that are not being used to convey their meanings or that are used as a quotation.

    e.g.   "Defective" and "faulty" can be used as synonyms.

    In the report conclusions, they wrote "will definitely."

- to give the translation of a foreign word(s), when the translation is placed between parentheses.

e.g.   The economist wanted to analyze whether the curve of the demand would remain stable ceteris paribus ("other things being equal").

- to recommend the use of some words.

    e.g.   I suggest that you write "asap" at the end of this letter.

    If one wants to be polite when writing formal letters in English, one should always preface negative or refusal statements with "I am afraid that".

- to write the plurals of italicized words.

    e.g.   The "dos" and "dont's" of metal fabrication technologies used within the automotive, aerospace, electronics and dentistry industries are going to be shortlisted.

- to use words ironically.

    e.g.   Non-expert audiences regard technical jargon as "jaw-breaking" terminology.

    The laboratory rats were not "in their quest for happiness."

- to use words from slang or very informal style if they are foreign to the writer or if they clash with the style of the rest of the text.

    e.g.   The manager being a "know-all," he chose that criterion to purchase the equipment.

    When writing to a non-expert audience, engineers usually try to avoid technical jargon as they know it is "gobbledygook" to lay readers.

- To mark the titles of books published within larger works, like chapters of books, names of articles, individual episodes in radio programmes, lectures or speeches.

    e.g.   "How to deliver a good oral presentation". (lecture)

    "Watching the Nanotube", IEEE Spectrum.

Single quotation marks (' ')

Single quotation marks are used:

- to enclose quotations within quotations and they are usually used after double quotation marks.

e.g.   "I am applying for a vacancy in the R+D department because coping with irate customers remarks like 'don't try to put me off' in the customer service department is more and more difficult for me", wrote Henry.

- to enclose a translation immediately following a foreign word or expressions, which is placed between commas.

e.g.   Doctors have lately realized that having a siesta, 'a nap', after lunch is a very healthy habit.
That car designer claims we should differentiate between genuine fin-de-siècle, 'stylish', design trends from post-modern tendencies.

INDIVIDUAL TASKS

5-22 The following excerpt has no punctuation marks. Provide as many punctuation marks as necessary to make the text more readable and correct.

sound cards for personal computers

a sound card is a conventional rectangular printed circuit board that is inserted into one of the free slots existing in the motherboard of a personal computer its main function is to convert data into an analog or continuous time signal which is later converted into sound by a pair of loudspeakers in the upper surface of the card there are two types of card interfaces that allow the user to connect other devices to the sound card these card interfaces are the jacks and the connectors the former could be divided into line in jacks line out jacks mic in jacks and speaker out jacks the mic in jack as its name indicates allows the user to connect the microphone that is used not only for recording voice but also for giving oral instructions to the computer the line out and the speaker out jacks perform the task of connecting external loudspeakers which amplify the signal coming from the sound card the latter connect the sound card to other devices such as a CDROM the loudspeaker of the personal computer or even an external joystick

in order to understand how the sound card processes data lets first consider which are the three main parameters that a sound card takes into account the sampling frequency 8KHz 22KHz 44 KHz the bit resolution 8 bits 16 bits 32 bits and the mode of working monophonic mode or stereophonic mode the process begins when data is sent from a storing device to a chip in the card known as DAC digital to analog converter this chip performs the conversion by receiving a discrete time signal in its input and delivering a continuous time signal to its output the higher the sampling frequency is the faster it works the higher the bit resolution is the better the quality the opposite takes place when recording with a microphone since now the processor has to convert an analog signal into a digital signal and therefore an ADC is used

5-23 Some punctuation marks have been omitted in the following letter. Read it and add those punctuation marks you think are missing.

Dear Ms. Jansen

I am currently a fourth year student at the University of Technology in Eindhoven and am interested in obtaining an internship position in the management accounting department at Shell as advertised on your web page.

Since 2001 I have been studying Industrial Engineering and Management. Most of the subjects in these studies deal with designing and optimizing all kinds of business processes like logistics, human resources and management accounting. In my last two years I specialized in the field of management account which is the reason why I am writing to you. During my university period I already heard and read much about Shell but when I joined a conference last summer in London I really became interested in your company. In fact what appeals to me most is not only the way of working presented at the conference but also the fact that Shell is a big international company.

The internship is about management accounting and takes at least six months I could start from the first of January next year needless to say I am highly motivated to work for you I am sure that my knowledge of management accounting could enable me to make a valuable contribution to your company.

I am enclosing my Curriculum Vitae. Please feel free to contact me for any additional information you may require. Thank you for your consideration.

Sincerely

Bart Beaumont

Enc Curriculum Vitae

5-24 The text below has been wrongly punctuated. Correct the punctuation where necessary.

Satellite uses and applications

In October 1957 when the UUSS amazed the whole world launching the first artificial satellite *the sputnik* a scaring and exciting race started involving the thirst for power of the two dominant world power states, however, what was conceived at the beginning as an expression of military and technological power has turned nowadays into one of the most common tools for communication, weather prediction, object positioning, or investigation. In fact the uses of satellites are so varied that one can find applications almost everywhere.
As mentioned in the previous paragraph. Satellites have many common and well-known applications, one of which is audiovisual communication. By simply placing a parabolic antenna on your roof (oriented towards the right position) you will receive the audiovisual

signal of countries placed; for instance, on the other side of the Atlantic Ocean. In this case, the satellite is placed in the *geostationary orbit* which allows a terrestrial receiver antenna to see always the same satellite in the same position. As a result, the receiver antenna captures the received signal, as a reflection from the satellite antenna. The fact is that the satellites placed at this stationary orbit are privileged not only because they move around the Earth at the same speed creating the sensation of a motionless satellite but also because this orbit is one of the most limited aerospatial resources, indeed, every country has its own section of geostationary terrestrial orbit, mostly addressed to audiovisual companies.

Satellites just as stars centuries ago have been found to be one of the most powerful positioning systems. Systems, like *gps* or *galileo* provide the user with one of the most accurate estimate of position. In contrast with other satellite applications, a whole system of satellite interaction is required to provide a positioning service available all the time. This system implies expensive constellations of hundreds of satellites, applications of this global positioning service range from the classical driving aid found in automobiles used by drivers to find the shortest way to airplane piloting, which is lately exploring the benefits of GPS, especially under bad weather conditions.

APPENDIX

KEY TO THE EXERCISES

Key to the exercises

This key aims to guide both teachers and students but it should not be taken as a closed, prescriptive answer key. On many occasions, we are providing one possible version of the task, which does not mean other ways of answering certain questions or approaching a given problem are not possible.

CHAPTER 1

TASK 1-1

Writing is necessary to transmit the merits of your work, your ideas and aspirations. It can be mostly improved by reading as much as you can and by asking a peer or colleague to review your writing.

Reading is necessary to improve your vocabulary, your grammar and your storytelling abilities; it is necessary to improve your writing. Reading can be improved by reading more.

Speaking is necessary to convey your ideas at meetings, to brief your peers after a business trip, to adapt a proposal to a client, to present a project to the general public, etc. Speaking can be improved by gathering as much information as possible to make you feel confident, by analyzing your audience's needs and accommodating your speech to them accordingly, by using PowerPoint or any other audiovisual aid with confidence, by practicing and rehearsing before your presentation, by sticking to the allotted time, and finally by looking for opportunities to speak.

Listening is necessary to communicate, whether in face-to-face interactions or over the telephone, for example. It can be improved by paying attention to what is said as well as to the speaker's intention, feelings and body language, by trying to reduce distractions, by looking at the speaker and responding appropriately (e.g. nodding), or by asking if you don't understand. You should always respect the speaker and never interrupt him/her.

TASK 1-2

The first text is certainly non-technical. There are many reasons that support this. If, first of all, we analyze the title, we realize that despite being somehow related to the technical field, it is not introducing a properly technical topic. If we go on reading, we come to the conclusion that the content is not technical due to the following reasons:

1) **Absence of technical vocabulary**. No technical expressions are used; instead we find some compound nouns, which are not really very technical (*information-carrying bits, savvy information-age citizen, hard-earned bits*). The acronym *IQ* appears without a definition as it is well-known.

2) **Absence of functions** such as definitions, description, instructions, etc.

3) **Absence of formulas, statistics, graphs and numerical expressions**.

4) **Use of informal style**. This is shown through some direct questions, personal pronouns (*I, you, we*) and informal expressions (*the right to bath daily in tubs of bits, watch the excess of information drain away, savvy*).

5) **Use of subjective tone**. Subjective tone is conveyed here with the use of expressions such as *I'm not sure, since we don't seem to have a lot of choice, when I no longer have the strength and ability...*

6) **Use of ironic tone**. From beginning to end an ironic tone is used and is reflected in expressions like *the era to come will be much better than the Ice Age, no one has told me what I'm supposed to do with all those bits, can I hide bits in my mattress, against the rainy day when I find myself bitless?*

The second text is clearly technical. It is technical because it displays many of the characteristics that define a technical text. As for the title, however, we cannot say that it introduces a technical topic by the way it is stated; it is too general to be taken as an early indication of the technical content of the text. Apart from the title, there are many more characteristics that corroborate the technicality of the text, among which we have:

1) **Use of technical vocabulary, compound nouns and acronyms**. There are many technical terms (*error burst, interleaving, interpolation, muting*), terms of Latin origin (*minimal, imperfection, erroneous*) and compound nouns (*compact disk digital audio, novel audio storage approach, digital storage technique, powerful error control coding scheme*) used all through the text. There are also two acronyms which are properly defined (*CD and CIRC*).

2) **Use of rhetorical functions**, specifically definitions of some of the technical terms previously mentioned (*interleaving, interpolation, muting*), function and process description in the last paragraph and finally visual-verbal relationship (*see Table 4*).

3) **Use of a numerical expression**: *two millimetres into an error burst of 2400 bits in length*.

4) **Use of a short list to organize information**. Some of the information in paragraph two is organized in a formatted list [(1) and (2)].

Finally, it should be noted that the tone is objective and straightforward as opposed to the subjective and ironic tone in the previous text.

TASK 1-3

1) Function description
2) Definition + physical description
3) Process description
4) Process Description (+ function description)
5) Instruction
6) Definition (+ function description)
7) Definition with physical description + classification
8) Physical description

CHAPTER 2

TASK 2-1

Text 1. *On Sprites and their Exotic Kin* is addressed to a *lay audience.*

a) The complexity of content is low. Note how the beginning of the text reminds us of that of fairy tales, which is a technique used to attract the readers' interest.
b) There are few sub-technical terms: *sprites*, *blue jets* and *elves.*
c) The sub-technical term *sprites* is defined.
d) A very clarifying picture of these optical emissions is provided.

Text 2. *Sources and Signals* is addressed to *students.*

a) The content of this text is rather more complex.
b) There are many technical terms: *transducer, message signal, waveform, baseband signal, analog signal, digital signal, amplitude, time, sampling, quantizing* and *encoding,* among others.
c) There are many definitions and examples. The terms defined include: *waveform, baseband, analog signal, digital signal, sampling, quantizing, encoding and sign bit.*
d) The terms clarified by means of examples are: *source of information, analog signals, digital signals, analog-to-digital conversion process.*
e) A block diagram, a figure and a table are provided.

TASK 2-2

		TEXT 1		TEXT 2		TEXT 3
Content	➢	**Technical**		Technical		Technical
		Sub-technical		Sub-technical	➢	**Sub-technical**
		General	➢	**General**		General
Language		Acronyms (non-defined)	➢	**Acronyms (non-defined)**		Acronyms (non-defined)
		Acronyms (defined)	➢	**Acronyms (defined)**		Acronyms (defined)
	➢	**Technical vocabulary (non-defined)**		Technical vocabulary (non-defined)	➢	**Technical vocabulary (non-defined)**
		Technical vocabulary (defined)		Technical vocabulary (defined)	➢	**Technical vocabulary (defined)**
	➢	**Compounds**	➢	**Compounds**	➢	**Compounds**
		Colloquialisms	➢	**Colloquialisms**		Colloquialisms
Format		Visuals		Visuals	➢	**Visuals**
	➢	**Numerical information**		Numerical information		Numerical information
AUDIENCE TYPE	➢	***HIGH-TECH***		*HIGH-TECH*		*HIGH-TECH*
		LOW-TECH		*LOW-TECH*	➢	***LOW-TECH***
		LAY	➢	***LAY***		*LAY*

TASK 2-4

The most difficult paragraph to write is the one addressed to a person who knows little about the topic. The reason for this must be found in the fact that writing to a layperson on a technical topic calls for a good command of writing skills. When writing to this audience, the writer must express ideas and facts clearly and cannot assume any implicit knowledge of the subject matter on the part of the reader, as when writing to technical experts. That is why texts addressed to a lay audience are often written by professional writers instead of technical experts.

TASK 2-9

Sample A: The purpose is *to ask for* some information.
Sample B: The purpose is *to inform*.
Sample C: The purpose is *to report*.

TASK 2-10

In keeping with this special issue on biometrics, this *paper presents* the facilities and network access-control applications of speaker-recognition and *focuses* on the three applications below.

TASK 2-11

Watching the Nanotube

This text is addressed to a *laypeople audience* and has the purpose of *informing* them about the current state of television technology (although the writer may also be indirectly persuading the reader). The writer tries to explain that plasma television, for many the last word in TVs, has in fact some important problems not yet solved (like image quality and high power demand) and that it has been outperformed by the latest technological advance in microelectronics, the nanotube. In particular, the writer seeks to set forth and highlight plasma's flaws and related business, commercial or governmental (i.e. electricity generation) implications while he stresses the most outstanding advantage of the nanotube TV from a consumers' wallet viewpoint—its lower consumption. However, although nanotube advantages clearly outweigh the plasma advantages, he doesn't explicitly ask readers to take action; he doesn't discourage them to buy a plasma TV, for example. Finally, the writer avoids using technical terminology (the acronym CRT is defined) and only defines nanotubes in passing and in a very casual way ('exotic molecules of carbon' and 'sheets of carbon atoms seamlessly *wrapped* into infinitesimal cylinders').

Facing a Difficult Problem

This text is written by a technical professional that addresses an executive in order to *inform* him about the current situation of Distronics and *ask for* some analysis and recommendation as some *action* is to be taken. The text (a memo) is written very to-the-point, straightforwardly and with clarity. Even though the writer is a technical expert in integrated circuits and other electronic components, he is hardly using any jargon or technical vocabulary. He knows he is addressing an economist or specialist in marketing who doesn't know much about electronics or integrated circuit technology. He uses a listing format to put forward the alternatives, a very useful technique if one wants to convey information orderly and concisely. This way, busy executives interested in the gist of documents can easily understand the message at a first glimpse and be in a better position to make decisions.

TASK 2-13

Audience and content.

TASK 2-14

In order to be persuasive, a leaflet to the general public written with the purpose of selling a product may well only emphasize its advantages. However, if you are writing to a well-informed expert from whom you cannot 'hide' information, your argumentation is probably going to be more credible, if you acknowledge and minimize the drawbacks or limitations the product may have before highlighting its benefits.

TASK 2-15

1) terminate(1), interrupt (2), stop (3), cut off (4).
2) telephone, ring up, make a call, give a buzz.
3) depart, fly, run away, do a runner.
4) collapse, failure, breakdown, crackup.
5) acquire, obtain, get hold of, score.
6) encounter accidentally, find by chance, run across, bump into.
7) thereafter, subsequently, afterwards, later.
8) yoke, collaborate, work together, team up.
9) abridge, diminish, reduce, cut down.
10) deplete, exhaust, finish, use up.
11) cease, conclude, finish, wind up.
12) be cognizant of, be aware of , know, be in the know of.

TASK 2-16

1) Do not disturb John; he's now concentrated on **solving** a mathematical problem.
2) My application for the internship in that audit-consulting company was **rejected** because, they said, I do not have enough experience in financial software.
3) The accountant assistant made a mistake when **writing** the cheque.
4) The After-Sales Department promised they would **carefully examine** our complaint and suggest a solution.
5) I **suggested/proposed** the idea of introducing a new payment system to keep the personnel more motivated, as the rate of absenteeism has been lately increasing. The Executive Board thought this suggestion was very good and accepted the proposal.
6) The alarm **was activated** when some burglars broke into the office building during the weekend.
7) After the terrorist attack, the manager decided to **make** a hundred workers **redundant** because business was very bad.
8) The company's plan to establish a sister company in Spain **failed** because the tax governmental policy is not appealing enough.
9) We have **interrupted** negotiations with this potential supplier.

10) The pipes in the system were leaking and the workshop manager (**re**)**solved** the problem by welding the loose joints.

TASK 2-17

1) They **obtained** good results and announced that they'd repeat the experiment with a different **method**.
2) The reply to our proposal was **somewhat** negative, but one of the **consequences** they said would happen was that our sales would **increase** for some time.
3) This year's benefits have been **very/rather** bad, but we'd better wait and see, **because** the political and economic situation of the country is **very optimistic**.
4) The new **employee** in the Quality Department has **spoiled** everything because he didn't have clear task guidelines
5) The factory workers seem to be **pleased with** the working conditions **for the present**.
6) We'll have to **contact** our suppliers to **resolve** what happened with those defective spares they sent.
7) The Executive Board will **get angry** as soon as the news that such an important client has been lost is known. However, the company shouldn't be **overly concerned with** this news because there are other important problems that need a solution **at the moment**.
8) The sales department can't **tolerate** delays in the assembly line as this only benefits the company's competitors.
9) These new materials have **been useful** for our research laboratory.
10) The purpose of this report is to present a **realistic** description of the **current** developments in the field of water supply and sewerage.

TASK 2-18

Contextual features		Situation One	Situation Two	Situation Three	Situation Four
No. of readers	1 or 2	√	√		
	Small group				√
	Large group			√	
Degree of Closeness	Known readers	√	√	√	√
	Unknown readers				
	Boss		√		

Status of participants	Peer				√
	Subordinate	√		√	
Shared background knowledge		√	√	√	√
	No				
Shared topical knowledge	Yes	√	√		√
	No			√	
Style		formal	formal	neutral	neutral-informal
Tone		business-like, personal	objective description of facts + a final subjective recommendation	personal, empathetic, assertive	irate, friendly

TASK 2-19

A.Formal style. Impersonal, cold and polite tone.
B. Neutral to informal style. Assertive, impolite and personal tone.
C .Neutral Style. Personal, helpful and polite tone.

TASK 2-21

Very formal: This research article sets forth a novel theoretic insight which will greatly impinge upon the field of microelectronics.
Formal: This research article proposes a novel theory that will become a major development in the field of microelectronics.
Informal: In this work you'll find an important theory that talks about microelectronics.
Very informal: Here's their last thing, kinda important stuff in microelectronics.

TASK 2-22

Dear Mr / Ms X,
I am one of the students of your Technical Written Communication class and I am writing to you to let you know I am afraid that I won't be able to attend the test. I have a compulsory laboratory session on Tuesday at the same time (10:00am). I wonder if it is possible to change the test from Tuesday morning to Tuesday afternoon.
Thanks in advance.
Best regards,
Mike Harris

TASK 2-24

Dear Paul,
I'm glad to let you know that the Faculty of Engineering has organized the Curriculum Project Forum and I'm personally inviting you to take part in it. As the Faculty Board wants to create an academic debate with a broad participation, they have suggested elaborating a project to help us face the future with an appropriate curriculum. Those of us who want to join the project are encouraged to raise relevant concerns, bring up new ideas, and send proposals.
If you're interested in collaborating, write to us, specifying in which of the four work teams you want to participate. Remember that the deadline for replies is March 12th.
Best wishes,
Ann Hurtley

(Note: depending on the relationship you have with your colleague, the tone and style may change considerably).

TASK 2-25

30 West Hampstead
London
January 20xx

Dear Tom Nolan,
It was a pleasure to hear from you again. We were very surprised not to have heard anything from you for such a long time. We hope this parenthesis was not due to any inconvenience caused by our firm.
As you will see in our brochure, we have introduced some important improvements in terms of design, specifications and price. We hope you appreciate this and, therefore, that you consider the possibility of placing an order with us.
We look forward to hearing from you.
Yours sincerely,
Jessica Nelson

TASK 2-26

It is safe to be cautious and somewhat tentative when putting forward results or drawing conclusions (particularly in science and technology, where the quick pace of advances and discoveries shorten their validity time). But a text with too many tentative markers would reflect a dubitative writer lacking self-confidence. This would give a negative image of the writer and his/her claims.

TASK 2-34

a) The outline can be improved in the following way: if the main idea is that pros outweigh cons and you want to conclude by saying that mobiles are indispensable, then you should look for at least one more advantage of mobiles. The improved outline could look as follows:

I.   Introduction
II.   Cons of mobile phones:
        A. need to be recharged
        B. more expensive than traditional phones
        C. loss of privacy
        D. exposure to microwaves: risk?
III.   Pros of mobile phones
        A. immediacy
        B. ease and comfort
        C. usefulness in emergency situations
        D. reachability
        E. multiple functions other than telephoning
IV. Conclusion: they have become an indispensable gadget

b) If the message you want to convey is that of an impartial observer, a listing approach can clearly reflect this idea.

I.   Introduction
II. Characteristics of mobile phones
        A. need to be recharged
        B. more expensive than traditional phones
        C. loss of privacy
        D. exposure to microwaves: risk
        E. immediacy
        F. ease and comfort
        G. availability
        H. reachability
        I. multiple functions other than telephoning
III. Conclusion: an objective description of these gadgets (no recommendation of evaluation)

TASK 2-35

Outline of the text *Stuff you don't learn in engineering school*:

I.   Introduction
        Recently graduated engineers have a deep technical knowledge but lack knowledge
        of non-technical skills: writing, speaking and listening communication skills.
II.   Writing skills
        II.a. Documents engineers need to write
        II.b. How to improve your writing:
                - use clear and direct language
                - read as much as you can
                - peer review
III.   Speaking skills
        III.a. Situations where engineers need to speak to a group of people
        III.b. You can learn by practising. Tips to improve:
                - control the topic (gather information, etc)
                - audience analysis
                - use Power Point proficiently
                - never read
                - stick to the allotted time
                - practice as much as possible
IV.   Listening skills
        IV.a. Difficult skill (many aspects come into play)
        IV.b. Tips to improve your listening skills:
                - reduce or eliminate distractions
                - make eye contact
                - respond appropriately
                - ask for clarification if you don't understand
                - pay attention and respect the speaker
V.   Conclusion
        Engineers need to have a good command of these to become effective engineers.
        Engineers are encouraged to be patient and practice and they will certainly become
        good writers, speakers and listeners.

TASK 2-36

Suggested outline on *The World Population Growth* based on a problem-solution approach:

I.   Introduction
II.   Description of the problem
        II.a. population growth rate in poor, underdeveloped countries:

 - fast growth of birth rate in poor countries: no contraceptive methods, lack of schools, hospitals & other governmental aid
 - fast growth of infant and adult death rate
 - children exploitation
 - lack of food and limited natural resources

II.b. population growth rate in rich, developed countries:
 - very low infant and adult death and birth rate: spoilt children, loneliness of elderly people
 - overestimation of materialistic values: a car is preferred to a child.

III.   Several solutions suggested (in underdeveloped countries)
III.a. Education:
 - increase literacy rate among women and children
 - increase literacy rate among the population in general
 - sensitize women to contraceptive methods

III.b. Health:
 - gain of a widespread sanitary aid: doctors, hospitals etc
 - decrease in the new born death rate

III.c. Economic support:
 - help these countries become economically autonomous
 - help these countries become richer
 - sensitize them to alternative sources of energy

IV.   Conclusions

Suggested outline on *The World Population Growth* based on a cause-and-effect approach emphasizing the negative effects:

I.   Introduction
II.   Main causes of the fast-growing population rate in poor underdeveloped countries
II.a. poverty
II.b. illiteracy
II.c. religious beliefs
II.d. absence of a welfare state
II.e. governments and politicians' passivity to take right measures
III.   Main effects of the fast-growing population rate
III.a. high infant birth rate in poor countries
 - many illiterate children
 - many ill-nourished and ill children (AIDS)
 - many exploited children
 - global earth population rate higher than ever before
III.b. pollution increase (worldwide)
 - more inhabitants
 - more cars, fridges, use of aerosols

      III.c. extinction of natural resources (worldwide)
         - water
         - food
         - fertile soil
IV.   Conclusions

CHAPTER 3

TASK 3-1

Paragraph 1: Since the Audio compact and CD-ROM were introduced ……….. In 1995 the successor to CD, DVD was announced.
Paragraph 2: Every DVD disc is made of ………. This disc could pack up to four movies (Normile 57).
Paragraph 3: Owing to the data capacity, …………., choosing from as many as eight different camera angles (Vizard 71).

Transitional sentence: In 1995 the successor to CD, DVD, was announced.
This transitional sentence can either be placed at the end of the first paragraph or at the beginning of the second.

Topics:
Paragraph 1: Capacity and limitations of the CD-ROM
Paragraph 2: Characteristics and configurations of the DVD
Paragraph 3: Advantages of the DVD

TASK 3-2

1) The future use of hydrogen cars
2) To protect computers from incoming viruses
3) Practising sports
4) Modern lifestyle
5) Driving too fast
6) The use of laser
7) The high emission of CO_2
8) Acceptance for publication
9) Effective ways to transmit information (using electric signals)
10) The invention of the chip

TASK 3-3

1) Bad (Statement of opinion)	6) Bad (Statement of fact)
2) Good	7) Good
3) Good	8) Good
4) Bad (Statement of fact)	9) Bad (Title)
5) Good	10) Good

TASK 3-4

The irrelevant sentence is the last one. However, this sentence could be considered a transitional sentence if another paragraph followed.

TASK 3-5

A) c   B) b   C) b

TASK 3-7

a) The explosion of the world's population during the second half of the 20th century was due to different reasons.
b) Computers have become essential tools in modern life because of their valuable characteristics. They...
c) The use of alternative sources of energy has several disadvantages.

TASK 3-8

1) The rapidity and effectiveness of maritime communications may be affected by several factors: hardware, lack of discipline in using a standardized procedure, and language barriers. Whereas the first factor depends on technology, the last two depend on human beings. The "human error" is found in a very large proportion of maritime casualties and often neither hardware nor the most advanced technology can eliminate it. Nevertheless, the losses in terms of lives and properties and the damage to the environment cannot be afforded, so we should work harder to reduce the risks associated with maritime disasters.

2) Satellites are essential in weather forecast. Weather forecast satellites use a rich variety of observations from which to analyse the current weather patterns. Since the launch of the first weather satellite in 1960, global observations have been possible, even in the remotest areas. Nowadays, it is possible to make a short-term weather forecast (5 days) and even a long-term forecast (3 months). This is extremely useful to predict cyclones, big storms and other catastrophes.

3) Background acoustic noise is a major impairment in voice mobile communications. The acoustic noise originates from the engine, air flow, frictions, and vibrations. It is coupled to the transmitted speech signal through the microphone, and therefore, deteriorates the signal to noise ratio at the transmitter. The problem is particularly significant in hands-free systems where the primary microphone is located far from the speaker. The effect of background noise is even more serious in integrated voice-data systems where the performance depends critically on the operation of speech detectors in a noise environment.

With permission from Groubran, R.A., R. Herbert & M.M. Hafez. (1990). Acoustic noise suppression using aggressive adaptive filtering. *Vehicular Technology* Conference, 8. © 1990 IEEE

TASK 3-9

As you may well have guessed, writers do not always include a topic sentence because the 'topic sentence technique' is not to be understood as a rigid and prescriptive rule, but rather as a helpful guideline. A text in which all paragraphs begin with a topic sentence, particularly a long one, will be too predictable and stiff, thus constraining writers' creative scope.

TASK 3-10

Sample 1. This is an example of a general-to-specific paragraph which begins by introducing a topic which is then developed. That is, the theme presented in the topic sentence, smog as an urban air pollution problem, is dealt with in the supporting sentences by describing its different types, analyzing its composition and stating its sources. Although it also resembles a problem-and-solution paragraph, as a problem is posed together with a discussion of the causes contributing to it, there is no identification of the solution. Therefore, the sample seems to follow a **general-to-specific pattern**.

Sample 2. This excerpt can be interpreted as following either a **problem-solution** pattern or a **specific-to-general** one. It can be said to follow a problem-solution pattern because the problem is described before a solution is suggested. On the other hand, it can also be seen as an example of a specific-to-general pattern in which different points are presented in order to draw a general conclusion. The author begins by describing the scarcity of information and inventories of plutonium and highly enriched uranium (HEU) in all areas, whether military or civil (specific info or problem description). Then he demands a greater transparency with respect to inventories of nuclear materials to help comply with and extend the Treaty on the Non-Proliferation of nuclear Weapons (generalization or solution).

Sample 3. The author of this passage is making some predictions as to the possible causes that may have caused a steel pot to leak. The results of the analyses seem to substantiate the hypothesis that the final wear and erosion of the pot is basically due to the presence of the large oxide defect in the steel. The passage, therefore, can be said to be based on **the effect-and cause pattern**.

Sample 4. The four main types of wear are listed in this passage according to the relative importance of their effects or costs. This way, the wear that cannot be eliminated, only reduced, is heading the list while the least costly or difficult to control is mentioned in the last place. The passage is based on a **listing in order of importance pattern**.

Sample 5. The four-stroke cycle of an internal combustion engine is described so that every cycle or stage is explained in the same sequential order the process takes place. The description of what happens at every stage leads us to conclude that we are facing a **sequential order pattern**.

Sample 6. This sample lists the four main types of data that should be gathered when describing the chemical properties of solid wastes. Therefore the paragraph is clearly based on a **listing in order of importance pattern**. In addition, as it includes a final evaluative comment on the importance of the chemical composition of such solid wastes, it can also be regarded as a **specific to general paragraph**.

TASK 3-11

1) First
2) Second / Next
3) Third /At this stage / Then / Afterwards…
4) Once / After
5) Finally

TASK 3-12

Since this paragraph follows a comparison-and-contrast pattern, the solution provided below is a possible way to improve the paragraph coherence in order to emphasize the point the author tries to make.

Filtration processes are used primarily to remove suspended particulate material from water and are one of the unit operations used in the production of potable water. There are two main types of filtration: cake filtration and depth filtration. These two processes are very different. Cake filtration is the physical removal by straining at the surface. With this type of filtration, **not only** filtrate quality improves as the filter run progresses **but also** deterioration of the filtered water quality is not observed at the end of the filter cycle. In addition, chemical pretreatments are not generally provided. **However**, to obtain reasonable filter cycles the source water must be of quite good quality. **On the other hand / In contrast**, depth filtration involves **more** complex mechanisms to achieve particulate removal. Transport mechanisms are needed to carry the small particles into contact with the surface of the individual filter grains, and then attachment mechanisms hold the particles to the surfaces. **Moreover**, chemical pre-treatment is essential to depth filtration. Rapid granular-bed filters are of the former type **while** precoat and slow sand filters are of the latter type.

TASK 3-15

1) comparison and contrast
2) chronological order + cause and effect
3) sequential order
4) problem and solution + cause and effect
5) general to specific + listing
6) comparison and contrast + listing
7) cause and effect
8) chronological order

TASK 3-17

1) problem and solution
2) comparison and contrast
3) chronological order
4) listing (exemplification)
5) cause and effect
6) sequential order
7) general to specific and/or listing
8) listing
9) listing in order of importance
10) analogy

TASK 3-18

The relationship established between these two sentences is that of causality and result even if no explicit connecting expression is used. It is the shared knowledge between writer and reader (it was the first time he did that so that's why he took the wrong tool) that unifies both sentences.

TASK 3-19

1) c		5) c
2) a		6) a
3) c		7) d
4) b		8) d

TASK 3-20

Passage 1: Any connector from the same group is also possible
1) Hence
2) However
3) Consequently
4) due to
5) such as
6) Although
7) that is
8) In spite of
9) Besides /Incidentally / Finally
10) in order

Passage 2:
1) therefore
2) First
3) thus
4) second
5) then /so
6) in the event of
7) since
8) not only

TASK 3-21

This passage has been made more coherent not only by adding several connecting expressions but also by using and/or repeating key words and pronouns.

GPS—a global navigation system everyone can use

Traditionally, navigation was an esoteric science until the US Department of Defense got fed up and said: "That's it! We've got to have a system that works". It was a massive undertaking. Since they had the money (over $12 billion) it took to do the system right, they came up with something to simplify accurate navigation, namely the Global Positioning System or GPS. This system is based on a constellation of 24 satellites orbiting the earth at a very high altitude. The satellites are high enough to avoid problems encountered by land-based systems and use technology accurate enough to give pinpoint positions anywhere in the world, 24 hours a day. Also, it is possible to get measurement accuracies better than the width of an average street. As GPS was first and foremost a defence system, it's been designed to be impervious to jamming and interference.

One of the most exciting aspects is its potential. With today's integrated circuit technology, GPS receivers are fast becoming small enough and cheap enough to be carried by anyone.

Everyone will have the ability to know exactly where one is, all the time. Knowing where you are is one of man's basic needs. As a result, this new service will become as basic as the telephone. In other words, it will become a 'new utility'. But that's just the start because its applications are almost limitless. GPS allows every square meter of the earth's surface to have a unique address. That means that new ways of organizing our work and play will be possible. Imagine a future in which /when a phone book is no longer a paper book, but rather a computer database in the memory of your computer. This database stores the exact GPS location of everything. For example, if you're looking for a Chinese restaurant, your computer can search through the phone database, find the location nearest to your current location, and direct you to it immediately. Thus, aimless hunting and wasted driving will be avoided.

TASK 3-22

(This is **one** possible version)

The microwave oven has many advantages and this is why it has become one of the great inventions of the 20th century. Millions of homes in America have one. Microwave ovens are popular because they cook food incredibly quickly. They are also extremely efficient in their use of electricity because a microwave oven heats only the food—nothing else.

The working principle for the microwave oven is *microwaves*, which, in fact, are radio waves. In the case of microwave ovens, the commonly used radio wave frequency is roughly 2,500 megahertz (2.5 gigahertz). Radio waves in this frequency range have an interesting property: they are absorbed by water, fats and sugars. When they are absorbed, they are converted directly into atomic motion—heat. Besides, microwaves in this frequency range have another interesting property: they are not absorbed by most plastics, glass or ceramics. Metal reflects microwaves and for this reason metal pans do not work well in a microwave oven.

Let's see how microwaves cook food. While in a conventional oven the heat has to migrate by conduction from the outside of the food toward the middle, in microwave cooking the radio waves penetrate the food. Radio waves excite water and fat molecules pretty much evenly throughout the food and so heat doesn't have to migrate toward the interior by conduction. Since there is heat everywhere all at once, the molecules are all excited together. However, there are some limits. First, radio waves penetrate unevenly in thick pieces of food, and second there are also "hot spots" caused by wave interference. To sum up, the whole heating process is different since in microwave ovens you are "exciting atoms" rather than "conducting heat".

TASK 3-23

1) Both a brief review of material behaviour and a description of the types of bar reinforcements are necessary.
2) You should apply a layer of this product and repeat the operation twice in order to protect the bar and to prevent it from getting rusted
3) The hinges squeak because they should be oiled and the cogs rub because they should be oiled too. Or: The hinges squeak and the cogs rub because both should be oiled.
4) After carefully evaluating the incident, we decided that we have neither the authority nor the means to cope with the problem.
5) This year many students have financial problems because university fees have risen and because fewer scholarships have been given by the government. / This year many students have financial problems because of the rise of university fees and the decrease of scholarships given by the government.
6) If you are to attend an interview, you had better wear smart formal clothes, arrive on time, and not slump into the chair.
7) As an engineer, you are expected to be a productive worker, a creative thinker, and an efficient communicator. / As an engineer, you are expected to work productively, think creatively, and communicate efficiently.
8) By taking action to thwart global warming, companies can not only reduce costs but (they can) also spark technological innovation.
9) Finding new sources of energy and developing economical means to convert natural sources of energy into usable energy are a major challenge in the 21st century.
10) The article sets out the reasons why business is taking global warming so seriously and explains how companies are preparing for a carbon-constrained world.
11) To combat climate change, scientists, governments and business must act fast, (they must) unleash the talent inside business, and (they must) pioneer technological innovation.
12) The lack of political will and public knowledge seem to be the main reasons for the inaction, but we have two choices—either we should be more concerned about global warming or more prepared to face the consequences.

TASK 3-25

Introduction 1: 5, 2, 1, 4, 3 / 5, 4, 1, 2, 3
Introduction 2: 6, 3, 4, 2, 1, 5

TASK 3-26

1) Incorrect.
2) Correct. Analogic and digital systems. Comparison and contrast.

3) Incorrect.
4) Correct. No subtopics mentioned. Comparison and contrast.
5) Correct. No outline mentioned. Chronological time order.
6) Correct. No subtopics mentioned. Listing.
7) Correct. A, B and C. Sequential order.
8) Correct: The hydrogen engine and the solar engine. Problem-Solution.
9) Incorrect.
10) Correct. Stress, sleep disorders and fatigue. Cause-Effect.

TASK 3-29

This is not a good introductory paragraph because the last sentence is not a thesis statement but a concluding sentence. A possible thesis statement could be, for example: *Satellites, then, can be said to have found a wide range of applications in many different fields.*

TASK 3-30

When the thesis statement anticipates the different subtopics and their pattern of organisation, readers can clearly foresee the information that will follow from the introduction. This can allow them to read more selectively and so discard in advance those subtopics which are not of their interest or decide that the subtopics are interesting enough for them to read the whole text. At the same time this makes the body of the text too predictable and may reduce the readers' interest. On the other hand, the fact that the thesis statement does not name the subtopics may arouse the readers' interest about the topic and how it will be developed. But we have to be careful in order not to reduce clarity and obscure the writer's informative purpose.

TASK 3-33

The final comment is a recommendation.

TASK 3-36

Nowadays, in developed western countries life is easier and more comfortable than ever before. If we want to see in the dark, go up ten floors, talk to somebody that lives far away, listen to our favourite rock band, or travel anywhere in a very short time, we only need to press a simple key. But not everything in modern life is advantageous *(as there is a price we are paying to enjoy these conveniences).*

There is a price we are paying to enjoy these conveniences. One of the problems derived from our lifestyle is that most of the devices referred to above work with electricity or petrol, sources of energy that are highly polluting. Another problem is that current energies such as petrol, gas, and carbon are non-renewable and will definitely be used up in a near future if we go on using them carelessly at the present pace. The day these energies are depleted, we should be ready to sacrifice our standard of living to a considerable extent.

For all these reasons, several initiatives are needed and actions should be taken both by the governments and by individuals to start looking for alternative energies that do not pollute the environment so much. For example, the Balear government will enforce a law that demands all new buildings incorporate solar panels. Similarly, if we want to keep using electricity, individual measures should be taken at a domestic level; simple actions like turning off the light and unplugging all appliances when no longer in use would certainly contribute to energy saving.

In summary, provided measures are taken by the State and provided we are seriously concerned and make an effort to be more respectful for our environment, I think that some of the problems related to the health of our planet may be partially solved.

TASK 3-37

This conclusion is incorrect because a new point has been added.

TASK 3-38

A reader well-acquainted with technical discourse will expect subjective opinions or interpretations near the end of the text or only after 'facts have spoken for themselves'. Such reader knows that, above all, technical discourse lends itself to a rational, empirical and objective exposition of facts that must come before the technical writer interprets them. Thus, this reader would possibly be on the alert, even distrust the writer, and think that the writer is trying to manipulate his/her opinion.

TASK 3-39

The essay is based on a listing pattern

TASK 3-40

Only topic (b) has been developed for you as an example.

Outline of an essay on vehicle safety systems based on a listing pattern:
I.   Introduction
II. Body paragraph 1: active safety systems: braking, steering, and suspension systems
III. Body paragraph 2: passive safety systems: seatbelt, airbag, steering wheel systems, pedestrian protection, safety electronics, tire pressure monitoring systems
IV. Conclusion

Outline of an essay on vehicle safety systems based on a cause-and-effect pattern:
I.   Introduction
II. Body paragraph 1: need to increase vehicle safety
III. Body paragraph 2: development of safety systems:
          - active safety systems enhance vehicle control and make driving safer.
          - passive safety systems reduce injury during collision or rollover.
IV. Conclusion

TASK 3-41

Outline 1: the outline is fine but it could be made more specific by adding supra categories. For example:
I. Introduction
II. Communicative aspects
          II.1.
          II.2.
III. Informative aspects
          III.1.
          III.2.
IV. Conclusion
Essay pattern: listing + comparison and contrast.

Outline 2: lack of a clear logical pattern of organization as ideas of different kinds are mixed.
Essay pattern: comparison and contrast (advantages and disadvantages), among others.

Outline 3: OK.
Essay pattern: chronological or time order pattern.

Outline 4: If we choose a problem-solution pattern, the outline could be improved by devoting one body paragraph to analyzing the problems caused by the Internet and another

body paragraph devoted to suggesting possible solutions. Alternatively, we could centre the essay on the three problems mentioned (i.e. pornography, internet at work, accessibility) and create three body paragraphs. In each paragraph we could put forward the problem and suggest solution(s).
Essay pattern: problem-solution.

Outline 5: this outline doesn't follow any pattern and thus could be greatly improved as information has been organized without following any logical criterion. In the first body paragraph we seem to find advantages, in the second we find that both pros and cons are dealt with, and in the third only negative aspects of the internet are tackled.
Essay pattern: depending on how you want to approach the topic, you can structure information following a cause-and-effect pattern, a comparison and contrast pattern, or a listing pattern, among other possibilities.

TASK 3-42

Introduction 1

I. Introduction (*Thesis statement:* E-waste is hardly recycled and has become a serious polluting agent.)
II. One reason why e-waste entails some hazard is that when old computers reach the end of their life, they are hardly recycled or dismantled (*topic sentence*). **(…)** The lack of e-waste recycling entails negative side effects (*transition sentence*).
III. The lack of e-waste recycling entails negative side effects (*transition sentence*). **One of these effects** (*transition*) arises from the polluting materials computers are mainly made from, which have proved to be harmful to mankind and environment (*topic sentence*).
IV. Conclusion

Introduction 2

I. Introduction (*Thesis Statement*: Intelligent houses incorporate some of the latest technological advances and environmental-friendly facilities not only to improve our comfort but also to contribute to our planet's sustainability.)
II. The first and possibly the most widely expected benefit of intelligent houses is that they are equipped with all sort of computer-centered mechanisms and user-friendly electronic devices that provide their occupants with the utmost comfort (*topic sentence*). (…) However, almost everyone in the 21st century would agree that a house fitted with modern comfort-giving facilities which are energetically inefficient is by no means the perfect dwelling (*transition sentence*).
III. **In effect**, (*transition*) the modern intelligent home must, without sacrificing comfort, be designed to save energy in order to obtain optimium efficiency using several methods (*topic sentence*).
IV. Conclusion

Introduction 3

I. Introduction (*Thesis statement*: This essay will analyze their main technical aspects— engines, suspension, brakes, and transmission.)

II. Let us begin by analyzing the first aspect concerning how the F1 engine and the WRC engine differ (*topic sentence*).

III. **As to suspension**, (*transition*) the F1 and the WRC cars have such different systems that it is difficult to establish a comparison (*topic sentence*).

IV. The engines and the suspension systems of these two types of cars are not the only striking differences to be singled out (*transition sentence*). These vehicles also require different brakes (*topic sentence*).

V. Finally, **(transition)** F1 cars and Rally cars are also unlike in one last important technical feature, their transmission (*topic sentence*).

VI. Conclusion

TASK 3-43

Changes made in terms of paragraph organization and topic sentences appear **in bold** whereas changes related to *inter-* and *intra-* paragraph coherence (connecting devices, unity, parallel structures, use of key words and pronouns, etc) have been underlined. (Note: This task has proved to be particularly interesting if students are asked to compare the key with their results, reflecting on their performance.)

Modern materials needs

Materials are probably more deep-seated in our culture than most of us realize. Transportation, housing, clothing, communication, recreation, and food production— virtually every segment of our everyday lives is influenced to one degree or another by materials. The development of many technologies that make our existence so comfortable has been intimately associated with the accessibility of suitable materials. An advancement in the understanding of a material type is often the forerunner to the stepwise progression of a technology, as in the case of sophisticated electronic devices and automobiles. Yet, even though tens of thousands of materials have evolved with rather specialized characteristics, our modern society needs increasingly more specialized new materials, most of which concern our planet's sustainability.

Energy is one of these current concerns. There is a recognized need to find new, economical sources of energy and, in addition, to use the present resources more efficiently. Materials can undoubtedly play a significant role in these developments. For example, the solar cells used to convert solar energy into electrical energy employ some rather complex and expensive materials. To ensure a viable technology, materials that are highly efficient in this conversion process must be developed so that they are even less costly. **Another example** is nuclear energy, which also holds some promise. However, the solutions to the many existing problems will necessarily involve materials, from fuels to containment structures to facilities for the disposal of radioactive waste.

Furthermore, research on pollution-free methods should be carried out following two directions. As it is known, environmental quality depends on our ability to control air and water pollution, so the more respectful and efficient the materials involved in many pollution control techniques, the better for the environment. Second, materials processing and refinement methods also need to be improved so that they produce less environmental degradation, that is, less pollution and less despoilage of the landscape from the mining of raw materials.

Transportation stands out as an additional field with ever-demanding new materials. Significant quantities of energy are involved in transportation. In order to enhance fuel efficiency, we can both reduce the weight of transportation vehicles (automobiles, aircraft, trains, etc) and increase engine operating temperatures. We need to develop new high-strength, low-density structural materials as well as (to develop) materials that have higher-temperature capabilities for use in engine components.

Finally, many materials that we use are derived from resources that are non-renewable; that is, not capable of being regenerated. These include polymers, from which the prime raw material is oil, and some metals. Given that these non-renewable resources are gradually becoming depleted, it is crucial to discover additional reserves and develop new materials having comparable properties and less environmental impact. The latter alternative is a major challenge for the materials scientist and engineer. Therefore, another significant concern points to the need of finding not only new materials as efficient as, or more efficient than, the current ones, but also of finding materials based on renewable resources.

Thus, in spite of all the tremendous progress that has been made in the understanding and development of materials within the past few years, there remain technological challenges requiring even more sophisticated and specialized materials that meet the needs of our modern and complex society. To make engineers' work more difficult, these new materials should cater for the current situation of our planet and the depletion of non-renewable resources. In a word /To sum up, new materials need being developed, with the caveat that they and their processing methods ensure the Earth's sustainability. We hope materials engineers will succeed in solving the problems we are faced with, as this would certainly be one important step in the technological progress that contributes to a better world.

Adapted from: *Materials Science and Engineering. An Introduction* by W. D. Callister. Copyright © 1994 John Wiley & Sons Inc., 2-5. Reprinted with permission of John Wiley & Sons, Inc.

TASK 3-45

As a matter of fact, some studies (Mauranen 1993, Valero-Garcés 1996, Dahl 2004, Hyland 2005, etc.) have proven that many of the inter-coherence expressions are used to guide readers while they are reading the text and that these devices basically orient readers by: connecting ideas, previewing content (i.e. expressions that anticipate and announce content that will appear later /below) and reviewing content (i.e. expressions referring to content that was already mentioned and that has a bearing at the time of writing). Since Spanish writers of English tend to underuse these expressions, and given that you are Spanish students, you should try to make an effort to be less writer-oriented and incorporate some of these expressions.

Students will provide their own examples, but to illustrate the point some of these expressions might be: *Likewise, note that X is another issue closely related to this analysis in particular; As we/you will see in the following section; As mentioned in the previous chapter; Remember the most important car features seen in the last section; Apart from collapse and loss of stability, the two causes described in the two sections above, there is one more important cause of uncertainty we cannot ignore —excessive local damage like cracking.*

TASK 3-46

	China	India
Science funding 2002-03 (US$ billions)	15.5	3.7
Workers in research and development	850,000	115,00
Doctorates produced per year	40,000	4,500
SCI-listed publications 2002-03	50,000	19,500
Percentage share of global publications	5	1.9

CHAPTER 4

TASK 4-1 Subject-verb agreement

1) The printer and the plotter we have bought **are** the best choice for us.
2) The new equipment is necessary to obtain the items of information that **are** going to buttress our research.
3) Gathering information without resorting to plagiarism **is** not easy in the age of internet.
4) The CD, together with the DVDs, **has** been stolen from the filing cabinet.
5) Everybody **knows** that the economics of the timber trade **is** ruling that country.
6) Versatility in engineers **has** proved to be highly appreciated among employers.

7) The analysis he made **was** rejected because the criteria he had chosen **were** neither specified nor homogeneous.
8) Provided the data **have/has** been corrected and the hypothesis of the work **has** been adapted to our goals, the government may sponsor the research.
9) The decision of using recyclable materials and of modernizing the coffee facilities in the office **has** been welcomed by the staff.
10) Frequent complaints **are** the battle that the quality department **has to** win.
11) What **needs** further analysis **is** the pitfalls of being computer-illiterate.
12) An appalling series of coincidences **has** apparently contributed to the success of these gadgets which **sell** so well.
13) The board **have not** been able to reach an agreement on how to fund the telecommunications programme, but either private companies or the government **is** likely to be the main sponsor. (In this particular example the emphasis is on the individual members of the board).
14) A sum of EUR 1000 **is** a lot of money.
15) The heating of the water is carried out without any direct contact with other liquids, which **prevents** clean water from being polluted.

TASK 4-2 Pronoun agreement

1) Neither of the two computers **has** been moved to **its** right place.
2) After two weeks of hard work, the team **has/have** announced that **it/they** will soon publish the results of **its/their** research.
3) The two technicians redesign the metal structure of the partition walls separately and then each **puts** forward **his/her** proposal for improvement.
4) The university should reconsider **its/their** performance evaluation system.
5) The Board receives complaints, analyzes them, and reaches **its** verdicts. Once reached, **these are** publicly announced.
6) The hypotheses and **their** corollary assumptions **have** affected the conclusions drawn.
7) There are many advantages and disadvantages of alternative sources of energy people should know and the media **need** to show some concern about **them**.
8) The committee will issue **its/their** yearly publication including all of the scholarship awards given.
9) One should try to introduce resourcefulness and eliminate monotony from **one's** work.

TASK 4-3 Clear pronoun antecedent

1) Even though Max is a very good product designer, he's never taken a lesson in that **subject/product design.**
2) Absenteeism was such a worrying fact that **the management** decided to penalize it./ Absenteeism was such a worrying fact that it was decided to be penalized.

3) The chief engineer Mr Higgins wished his computer had saved the information in **its** memory before the lights went off.

4) **The research article claimed** that the latest fatigue resistant composite had in fact negative long-term consequences.

5) Clara's colleague told her **that Clara/that Clara's colleague** shouldn't have refused the promotion.

6) A leg of a robot was thrust into the air, hitting the microscope and breaking the test tubes next to **the microscope**. Strange though it may seem, **the microscope/the leg of the robot** was only slightly damaged.

7) The two civil engineers, acting on behalf of their company, made a formal complaint about the irregularities in the tunnel construction. **This complaint/These irregularities** irritated the mayor.

8) The two tunnel construction methods have been widely used around the world. The first is said to be cheaper and only slightly less safe than the second. Therefore, it is not clear why the builder cannot choose **the first/the second**.

9) The first ever-created multi-disciplinary research group of doctors, biologists and engineers working on genetics is hovering on the brink of a new breakthrough. **This research group/ breakthrough** is expected to benefit humankind.

TASK 4-4 Subject repetition

1) It is believed that **the alarm can be triggered** simply because of thunder.

2) **The system is described** in the first part of chapter and in the remainder statistical analyses are performed.

3) It is feasible to build the new laboratory this year, as **the construction is financed** by the autonomous government.

4) Recently, **studies (have been performed)** whose aim was to obtain analytic expressions that allow the study of the rectifier behaviour in an easy way **(have been performed)**.

5) As soon as **the chip is installed**, the unit is expected to operate properly again.

6) In this central heating system in particular, **the pump is also very important** because it makes water flow through the pipes and radiators.

7) **Twenty-one vacancies** in the department **were estimated**. / **It was estimated that** there would be twenty-one vacancies in the department

8) The university researchers have been given permission to carry out the research project which ~~it~~ will be sponsored by the local administration.

9) **The samples were analyzed** and the result that ~~it~~ was obtained from the statistical analysis was checked.

10) Finding alternative sources of energy and familiarizing the population with them **is** not an easy job.

TASK 4-5 Subject omission

1) **There** exist two alternative factory layout designs (Or: Two alternative factory layout designs exist) but we are waiting for the environmental engineers to issue their feasibility report.
2) **It** has been proved that the chemical factory is disposing of hazardous waste and throwing it into the river. (Or: The chemical factory has been proved to be dispensing...).
3) The seatbelt is part of the safety equipment in a car; if well used, **it** is important because **it** can prevent you from being seriously injured in a car accident.
4) Since the pipes are usually made of metal and since water can produce corrosion on them, **it** is advisable to check the installation once a year (Or: The installation had better be checked...).
5) It is known that **there** exists the problem of keeping all rooms in a house at the same temperature.
6) Hold the mouse with your right hand. As you hold it, **it** is necessary for you to look at the monitor while you are moving the small device.
7) Installing a door switch alarm system at one's home is a good idea. If a thief broke into the house, **he/she** would activate the alarm and **it** is very likely that he/she was caught by the police.
8) Nowadays **an important problem is arising** in our planet: the continuous exploitation of natural resources has caused an increase in pollution and an alarming shortage of these resources.
9) In the process of selecting the parts of a water pumping station **there** are two options—a diesel engine or a gasoline engine.
10) With regard to fuel cells as a type of power technology, **it** is hoped that efficiencies of over fifty percent can be achieved in the next ten years.

TASK 4-6 Dangling modifiers

1) Writing a memo to his superior, **he made** a mistake concerning the total amount of money.
2) Towed behind the tug, **the oil tanker** entered the harbour.
3) When writing a document, **one/you** must take care not to misplace modifiers.
4) An Integrated Circuit (IC) is bonded to the leapframe plate using either a eutectic solder or an epoxy resin, **materials typically used for the IC.**
5) Without a mobile phone, **nobody** could communicate in that area.
6) If the air-conditioning device is to be installed, the technician should make a hole **that allows cooling the room** in the wall.
7) Nobody realized that the mechanisms **checked by the quality manager** were not working properly.
8) The first table **in the results section of the article** has to be proofread.

9) Nobody knew how to mend the video camera**(, which had been) broken** by some angry workers.
10) Ignoring the price, **they** bought the latest model of digital cameras.
11) After analyzing the method used, **the scientist** realized that the results were unconvincing./ After analyzing the method used, **we** found the results unconvincing.
12) The experiment failed **because we hadn't followed** the instructions carefully.
13) The principal types of operation **used to work laminated plastic sheets** are: sawing, drilling, routing, beveling, bonding, and forming.

TASK 4-7 Reduced relative clauses.

1) The drilling machinery **finally installed** costs EUR 12 300.
2) One of the moons **revolving** around the planet was photographed.
3) People **living** near airports often suffer from heart problems **caused** by noise pollution.
4) The island **destroyed** by the tsunami is being rebuilt thanks to international help.
5) Candidates **interested** in the vacancy may submit their curricula vitae (,**which should arrive**) by the end of the week.
6) Pieces of equipment **requiring** intensive care carry a 3-year guarantee./Pieces of equipment **carrying** a 3-year guarantee require intensive care.
7) The printer, **overused** by the staff, must be urgently replaced./The printer **overused** by the staff must be urgently replaced.
8) A new research laboratory **fitted with** the most advanced technological equipment is being built in the campus.
9) The **leaking** pipe was repaired overnight. /The pipe, **leaking** for hours, was repaired overnight.
10) The installation, **completed** only a year ago, already needs repairs.
11) The reports, **reviewed** by my superior, will be discussed tomorrow.
12) Nowadays libraries are subscribed to many journals, **which may offer** only an electronic format (The clause is not reduced in order not to lose the extra meaning of the modal *may*).

TASK 4-8 Run-on sentences

1) Three hundred immigrant workers worked non stop; **as a result**, the five-star spa hotel was built in about two years./ **Because** three hundred immigrants (...), the 5-star spa hotel (...).
2) A zinc coating has been proved to reduce resistance; **therefore**, it is no longer used in vehicles nowadays.
3) The frame of mountain bikes has not changed very much over the years, **but** in the last few years frames have increasingly been made of lighter materials like aluminium.
4) Engineering students used to make two-dimensional drawings by hand. Nowadays engineering students make three-dimensional drawings with the help of CAD programs.

5) Peter's father approves of his son's decision. **However,** I know that underneath he would like his son to follow him into the engineering business.

6) Poor quality and lack of rigour angers my boss, **who /as he** believes these attitudes have permeated the system./Poor quality and lack of rigour angers my boss. He believes these attitudes have permeated the system.

7) The company is about to launch a new protective foam in compliance with the European regulation and meeting all European safety standards. **(Consequently,)This** is expected to be the money-making product of the company.

8) The new legislation offers EUR 2 billion to finance solar energy **projects; as a result,** the number of companies interested in these projects is expected to boom.

9) A newly issued report reveals in facts and figures what the average citizen has known for years: /**; namely,** the inflation rate has been steadily increasing since the Euro first appeared.

10) You should try to proofread all your documents for run-on sentences and sentence fragments. **You** should also proofread them for any other grammatical error.

11) Erasmus students willing to access the intranet of virtual courses should enter their user's name, which is their identity card number, **and** their password, which is their date of birth plus two more digits. **Then they should** click on elective virtual courses and enter the intranet. **There** they will find exercises, texts and other documentation together with the teacher's instructions.

12) You had better do the homework and send it via intranet before the deadline. **Afterwards,** you will receive an acknowledgement of the receipt from the teacher.

TASK 4-9 Sentence fragments

1) High achievers are usually ambitious people **who** also like taking risks.

2) The engineering academic identified three main stressors among his students: exams, lab reports and essays, and oral presentations. (Or: , **namely**)

3) Women engineers do not always have the same career opportunities as men engineers (,) even though private and public companies do not always accept this reality.

4) Delving into the cause-effect analysis of the problem, they realized they had initially underpinned loss of ductility as a side effect while in fact it was a cause.

5) One of the greatest engineering works undertaken in Europe, this suspended bridge was built in a record time of **less** than 2 years. /... was built in a record time, **less** than two years.

6) Following her lawyer's advice, Laura decided it was worth spending time and money and patented her invention last year.

7) Ms Edwards was initially reluctant to change her department organigram but she was finally convinced by Jim Leeds, a highly reputed professional. Or: Ms Edwards, **a highly reputed professional,** was initially reluctant to change her department organigram but she was finally convinced by Jim Leeds.

8) These laptop computers stand out as the cheapest and most reliable in the market since they have been manufactured at a very low cost near Shanghai.
9) Owing to the increase in scholarships, a greater amount of students have applied for economic help this year.
10) The economist announced that Singapore, India and China could be singled out as three emerging economic forces, a widely accepted opinion.
11) If the chief computer engineer had had no fear of more viruses invading the company's net, no special measures would have been taken.
12) The architect is known to be eccentric and excessively ambitious among his co-workers. However, **he is** considered a genius.

TASK 4-10 Parallel structures (revision)

What is Biotechnology?

While biotechnology has been designed in many forms, in essence, it implies the use of microbial, animal or plant cells or enzymes to synthesise, (to) breakdown or (to) **transform materials**. The aims of the European Federation of Biotechnology (EFB) are: i) to advance biotechnology for the public health; ii) **to promote awareness, communication and collaboration** in all the fields of biotechnology; iii) to provide governmental and supranational bodies with information and informed opinions on biotechnology; iv) to promote public understanding of biotechnology.

The EFB definition is applicable to **both** 'traditional or old' biotechnology **and** 'new or modern' biotechnology. 'Traditional' biotechnology refers to the conventional techniques that have been used for many centuries to produce beer, wine, cheese and many other foods **while 'new' biotechnology refers** to the methods of genetic modification by recombinant DNA and cell fusion techniques, together with the modern development of traditional biotechnology. The difficulties of defining biotechnology and the resulting misunderstandings have led some people to abandon the term 'biotechnology' as too general **and to replace it by the precise term** of whatever specific technology was being used.

Glass Fibers

Glass fibers are unique materials that exhibit the familiar bulk glass properties of hardness, transparency, **resistance** to chemical attack, stability, and inertness, as well as fiber properties of strength, flexibility, **lightness of weight**, and processability. Glass is made by fusing silicates with **soda, potash, lime, and various metallic oxides**.

Two basic processes are used to manufacture continuous glass filaments: marble melt and direct melt. In the marble melt process, glass marbles are first produced by melting the appropriate mixture of raw materials and forming marbles, which are usually 2 to 3 cm in diameter. These marbles are then remelted at the same or different location **and formed into** the glass fiber product. **In the direct melt process, the raw materials are melted and**

formed directly into the glass fiber product. Direct melt is the primary process used in continuous glass fiber manufacture today.

TASK 4-11

Passage 1
The history of Sheetco may be useful to illustrate this point (1). This year, the manager of the company, manufacturer of special tools and presses necessary to work with laminated plastic sheets like Formica, **has decided** (*subject-vb agreement*) to acquire a new installation for the two product lines of the company and improve the customer service department (2). **It** (*Subject omission*) has become necessary to reduce the company's distribution channels with all these changes (3). In order to improve the first product line, one needs to understand the manufacturing process of the sheets that Sheetco produces (4). First, the laminated plastic sheets are made of specially processed papers impregnated with resins (5); then these sheets, or layers, are cured under intense heat and pressure (6). The layers ~~which~~ (*sentence fragment*) are subsequently fused into sheets usually about 1/16 inch thick (7). The end product is sheets **bonded to plywood and hardboard** (*Dangling modifier*) that can be used as a surface material, on many types of domestic, commercial and industrial furniture and furnishings (8). An example of one of the most popular applications of these sheets in the home is to be found in kitchen counter tops (9). As to the second line, Sheetco also manufactures their own bonding and forming presses and sells them, as said above (10). The 'Plus Vacuum' press, **the Company's best press** *(dangling modifier),* is based on a heat process which saves a great deal of time over the conventional cold-pressure method of bonding plastic sheets to a second surface (11).

Passage 2
Many of our modern technologies require materials with unusual combinations of properties that cannot be met by the conventional metal alloys, ceramics, and polymeric materials (1). This is especially true for materials **needed** (*reduced relative clause*) for aerospace, underwater, and transportation applications (2). Aircraft engineers are increasingly searching for structural materials that ~~they~~ (*double subject*) have strength, stiffness, low densities, and corrosion, abrasion and impact resistance, for example (3). **This** (*clear pronoun antecedent*) is a rather formidable combination of characteristics (4)**. Because** (*run-on sentence*) strong materials are frequently relatively dense, increasing the strength or stiffness generally results in a decrease in impact strength (5).

Passage 3
As you know, the study of a circuit building block, the operational amplifier (op amp) is of universal importance (1). Early op amps were constructed from discrete components (vacuum tubes and then transistors and resistors) and **their** (*pronoun agreement*) cost was prohibitively high (2). In the mid-1960s the first integrated circuit (IC) op amp was produced (3). An IC op amp is made up of a large number of transistors, resistors, and (sometimes) a capacitor connected in a rather complex circuit (4). Although its characteristics were poor by

today's standards and its price was still quite high (5) (*sentence fragment*), **its** appearance signalled a new era in electronic circuit design (6). Electronics engineers started using op amps **in large quantities** (*dangling modifier*), **which it** (*double subject*) caused their price to drop dramatically (7). They also demanded better-quality op amps (8). Semiconductor manufacturers responded quickly and within the span of a few years high-quality op amps became available at extremely low prices (9).

In order to take proper advantage of the economics of integrated circuits, designers have had to overcome some serious device limitations (such as poor resistor tolerances) while exploiting device advantages (such as good component matching)**; therefore,** (*run-on sentence*) an understanding of device characteristics is essential in designing good integrated circuits (10). Also **it is** (*subject omission*) helpful when applying commercial integrated circuits to system design (11). Germanium and gallium arsenide are also used to make semiconducting devices**; however,** (*run-on sentence*) silicon is still the most popular material and will remain so for some time (12). The physical properties of silicon **make** (*subject-vb agreement*) it suitable for fabricating active devices with good electrical characteristics (13). In addition, silicon can easily be oxidised to form an **insulating layer** (glass) (*reduced relative clause*) (14). **It has been proved** (*clear pronoun antecedent*) that this insulator is used to make capacitor structures and allows the construction of field-controlled devices (15). It also serves as a good mask against foreign impurities, which could diffuse into the high-purity silicon material (16). This masking property allows the formation of integrated circuits; active and passive circuit elements can be built together on the same piece of material (substrate) at the same time and they can be interconnected to form a complete circuit function (17).

TASK 4-12

1. We have also learnt that the technical skills of an engineer who has a postgraduate degree from a highly-reputed foreign university are very useful to the development of his/her professional career. This is especially true if this degree focuses on the specific area the company needs additional resources for which, more often that not, is not the case.

2. The first conclusion we reached was that, additionally to the deep and sound knowledge about engineering, the most suitable candidate should have some special abilities, namely social and communication skills. These skills are absolutely necessary for an engineer working in a Production Department. However, the research concluded that the experience is only important as far as it helps candidates to improve the skills mentioned above.

3. Air pollutants are either *gaseous* or *particulate*. Some common gaseous pollutants include sulfur oxide (...).Common particulate pollutants include cement dust (...). Particulate emissions like smoke are typically easier (...).

4. I am writing to let you know that there are problems that I still need to solve to finish my project. The main problem is that I don't know yet what kind of transmission I am going to use. Without this information I cannot decide on the car's dimensions and, consequently, my work in the project is being delayed.

5. Living in a dormitory is more uncomfortable but more interesting than living at home. The final decision will depend on the kind of person you are or you want to be. This decision also gives you a good opportunity on whether you want the easier way of life (i.e. living with your parents) or on whether you think it is time to learn about how difficult life is. This will also help you become an adult, mature person.

TASK 4-13

1) Because the employee was tired of waiting for a promotion, he accepted a new job, which was 50 miles from his house. *Or*: Tired of waiting for a promotion, the employee accepted a new job, which was 50 miles from his house.

2) The conference on microelectronics was given by a well-known expert who had been awarded an important prize because of the important discoveries he had made in this field. Consequently, the conference had a massive assistance.

3) Speech-recognition devices are data entry devices that can recognize the words spoken by a person. However, they are expensive devices that are not mass-produced.

4) Even though the power was very slow and the processor could not work at its normal speed, we could still manage to finish the work.

5) To prepare the job interview, you must work hard and consider all possible questions about you and the enterprise; otherwise, you may have trouble finding the proper answers.

6) I told Eric to wait in my office while I was photocopying, in the event the phone might ring.

7) The GPRS is a wonderful tool that enables customers to use their phones for services other than voice. For example, customers will be able to access their company's intranet, book flights, check the weather or find out what's on TV.

8) The building showed suspicious scratches; therefore, they decided to repair it because they wanted to sell it and buy a new one.

9) A group of specialists studied the latest and most relevant information collected by a research group, and yet could not reach a satisfying conclusion.

10) A CD is a laser-read data storage device which can store audio, video and textual material. Unlike a conventional tape, a CD stores information in digital, hence its absence of background noise.

TASK 4-14

1) unjustified, 2) justified: redundant agent, 3) justified: unknown agent, 4) justified: not to be offensive 5) unjustified, 6) unjustified, 7) justified: impersonal, 8) justified: unknown agent (probably a research team), 9) unjustified, 10) unjustified

TASK 4-15

Radioactive or nuclear waste is the unusual byproduct of nuclear activities. These wastes are classified as high-level waste (HLW), transuranic (TRU) waste, or low-level waste (LLW). Each of these types of nuclear wastes presents its own challenges for proper management.

High-level waste (HLW) is the highly radioactive waste resulting from reprocessing spent fuel, which is the fuel that has been irradiated in a nuclear reactor. Currently, in the U.S., reprocessing involves removing the plutonium and uranium from spent fuel **that the USDOE nuclear reactors generate**. The plutonium and uranium are recycled for use in defence programs. What remains after reprocessing is highly radioactive waste that must be remotely handled behind heavy protective shielding. **This waste requires long-term isolation**, typically in an underground repository, while it stabilizes. Shipping HLW presents a unique hazard, so it is packaged in heavily shielded containers for storage and transport.

The spent fuel resulting from the generation of electricity from U.S commercial nuclear power plants is currently not being reprocessed. The Nuclear Waste Policy Act of 1982 is the federal statute that creates the framework for managing nuclear waste. Under this act, the nuclear wastes from nuclear power plants are to be placed in an underground repository for long-term isolation. **The act set forth ambitious deadlines** for siting, constructing, and operating two geologic repositories. Five years later, **the Congress passed the Nuclear Waste Policy Amendments Act of 1987**. These amendments built on the previous direction of the underground repository program and in fact required that DOE focus on one site -- Yucca Mountain in Nevada. Two repositories are currently being developed –Waste Isolation Pilot Plant (WIPPs) in New Mexico for waste generated by the DOE, and Yucca Mountain for waste generated by commercial power plants.

TASK 4-16

The text includes **jargon**, sub-technical vocabulary, _Latin terms_, *acronyms* and **abbreviations**.

The wide **bandwidth** of *UWB* pulses _precludes_ the _use_ of traditional _receiver architectures_. Instead a simple sensor has been _employed_ that detects energy that is _instantaneously present_ over a bandwidth of several **GHz**.

The sensor fulfils three _functions_; namely, _detection, classification_ and _geo-location_. A high _gain_, **spinning _antenna_** is used for detection and will also provide better reception for classification. An *IRA* has been _developed_ to best meet these _requirements_. **Broadband**, _**omni-directional** antennas_ are used for the three **geo-_location channels_**.

The electronics of the sensor _consists_ of a detection _system_ and a **broadband collection system**. The detector has been _designed_ to detect *UWB* _pulses_ in the presence of _interference_ and _provide_ a trigger to the _collection_ system. High speed *ECL* **logic** has been used to _minimize propagation_ delays through the detector. A Tektronix TDS744404 with an **analogue input bandwidth** of 4**GHz** has been used as the collection system. One channel is used to collect _data_ from the *IRA* and the other three channels are used to collect data from the omni-directional antennas.

TASK 4-17

1) Tell the secretary to take an **extra** copy in case the general manager's **forgetful PA (Personal Assistant***)*** fails to remember.
2) The **improvements** applied to the **beam** didn't guarantee its bending strength (i.e., the beam's resistance to bending moments).
3) The **first** control **brought about** important information and **set up** the basis for **next** controls.
4) Some of the more popular *AI* **(Artificial Intelligence***)*** methods used in data **filtering** include neural networks, clustering and decision trees.
5) The client showed an **apparent** disagreement **about** the way the **business** was **carried out**.
6) **PCBs *(Printed Circuit Boards)***, circuits and other electrical devices were all placed in large boxes with the purpose of **starting** the change of laboratory **as soon as possible**.
7) If you should **buy** a pen drive, make sure it has **at least** a capacity of 1Gb *(**Gigabyte***)*.

TASK 4-18

1) The **debugging** of the **initial** program was a **complex task/ undertaking**.
2) Ground-wave propagation is the **dominant mode** of propagation for frequencies in the **FM** band. This is the frequency band used for **AM** broadcasting and **maritime** radio broadcasting.

3) They **endeavoured** to protect the program from being **plagiarized**.
4) The bolt weakened because of **corrosion**, causing the **crankshaft** to break.
5) Another major **advantage** of *CAD* is that it can **interface with** other programs.
6) The **initial** step in a radio system is **to select** the required **RF** wave from those **received** by the aerial.
7) I hope you **realize/appreciate** the **measures** that the trainee engineer has proposed to **optimize** the tasks on **the assembly line**.
8) With Apple's **iPod,** anybody can now **transfer a large number of** songs into this device at **affordable** prices.
9) More powerful and faster **PDAs** are coming out **at the moment**. The latest models **incorporate/ contain** a **GPS**.

Completing the lists: Redundancy and wordiness (pp. 215-216)

Redundant	**Direct**
Basic essentials	Essentials
Consensus of opinion	Consensus
Mutual cooperation	Cooperation
End result	Result
Utmost perfection	Perfection
Local neighbourhood	Neighbourhood
Different varieties	Varieties
Physical size	Size
Triangular in shape	Triangular
Uniformly consistent	Consistent
First introduction	Introduction
On a daily basis	Daily
New innovation	Innovation
The month of July	July
Actual experience	Experience
Final result	Result

Wordy phrases	**Shorter expressions**
The majority of	Most
In close proximity	Near
Readily apparent	Obvious
Prior to	Before
In the course of	During
With reference to	About
Exhibit the ability	Can
In the forms of	As

Wordy phrases	Shorter expressions
A number of	Some
In accordance with	by, with
In the event that	If
With the exception of	Except for
Provide guidance for	Guide
In the near future	Soon
In conjunction with	With
Put one in place of another	Replace
To one exceptional degree	To one extent

TASK 4-19

1) The problem was solved by applying the Fourier transformer.
2) First the signal is received and then it is processed.
3) Hydrogen cars would greatly reduce air pollution, especially in big cities.
4) Printing quality increases with a laser printer / A laser printer increases printing quality.
5) The two parties met in the human resources department and they finally agreed on the mode of payment.
6) I was baffled when I realized she knew a lot about the programming language and she solved tasks in an efficient manner.
7) To launch the new product first we analyzed the market needs, then we set the objectives and finally we presented the product.
8) Although he gained on-the-job training with the internship, he left the company alleging mobbing.
9) The maintenance department carries out yearly tests and controls within the safety program.
10) Replacing the hard disk will cost about EUR 120.
11) The Environment Quality Agency has issued his assessment outcome, which reveals that we need to encourage environmental technology and raise awareness among the general public.

TASK 4-20

1) Every secretary should have reported to **his /her** boss at the end of the day / **Secretaries** should have reported to **their** boss at the end of the day.
2) If a **worker** spots any malfunctioning in the machinery, **he/she** is expected to warn the **supervisor** at once / If **workers** spot..., **they** are expected to warn...
3) A **police officer** requisitioned a cargo of **hand-made** earthenware that was being illegally introduced into the country.

4) The chief doctor in the department is responsible for all doctors and **nurses**. **His/Her** duties include organizing tasks and controlling work, among other things. / The chief doctor, Dr **George Roberts**, is responsible for all doctors and **nurses**. **His** duties include organising tasks and controlling work, among other things.

5) Some sparks are thought to have caused the small fire in the assembly plant. A very **efficient** employee witnessed what happened and warned the secretaries. **These** quickly called the **firefighters** and the fire was soon extinguished.

6) If an applicant for the electronics engineer post wants to be seriously considered, **he/she** will have to hand in an achievement-based curriculum./If **applicants**..., **they**...

REVISING AN ACADEMIC ESSAY: FINAL DRAFT

Hydrogen: the future is coming

The lifestyle and standard of living in developed countries and the industrialization in emerging developed countries such as China and India are causing fossil fuel supplies to run out. The need to find a new combustible, which allows us to keep our lifestyle, can be satisfied thanks to the use of hydrogen. This new energy source can be a great solution given that it is an endless and clean resource with a great variety of uses. First we are going to compare hydrogen with fuel cells and then we will see the applications and the advantages of the hydrogen cell.

In principle, a hydrogen cell operates like a battery: it produces energy in the form of electricity and, as a result of some undesired losses also heat, as long as fuel is supplied. A fuel cell consists of two electrodes sandwiched around an electrolyte. Oxygen passes over one electrode and hydrogen over the other. As long as fuel and oxygen are supplied to the cell, this will keep producing electricity for ever. The oxygen needed by a fuel cell is usually simply obtained from air. Hydrogen cells can be thought of as devices that do the reverse operation of that in the well-known experiment where an electric current passing through water splits it up into hydrogen and oxygen. They can be used to produce from quite small amounts of electric power, for devices such as portable computers or radio transmitters, to very high power ratings for electric power stations.

In fact, the hydrogen cell has a lot of applications. Not only is hydrogen used in cars but also in airplanes and even in some household appliances like electric scales. Nevertheless, The development of the cell is focused on cars since they are the main future market. Some of the giant car companies are also designing hydrogen-powered cars. For example, BMW is going to sell in 2007 one hundred cars whose fuel is not petrol but hydrogen. Furthermore, British Petroleum (BP) has opened a hydrogen station in Beijing in order to encourage people to use hydrogen-powered cars. Finally, governments are also financing projects that investigate how to make a better use of hydrogen cells and are sponsoring initiatives that involve the use of hydrogen. This way, cars which use renewable sources will be tax-free.

Another interesting point to be mentioned is the differences between petrol and hydrogen cars in order to better understand the advantages of one over the other. On the one hand, petrol cars have both a better mechanical efficiency and higher power. On the other hand, hydrogen is an endless and renewable source of energy; besides, as it does not produce polluting emissions, it does not damage the ozone layer. Despite the fact that hydrogen cars are less powerful than fuel combustible ones, BMW has managed to manufacture a car with 285 HP that accelerates from 0 to 100 Km/h in six seconds and reaches a maximum velocity of 302 Km/h. It is a great example of how the latest projects can give benefits to companies. In conclusion, we could say that hydrogen is a good solution to resolve the shortage of fossil fuel sources and to reduce contamination. By means of a simple chemical process, we are able to obtain hydrogen from water, which at the moment is an abundant resource in nature. We engineers are able to obtain a clean energy source which can be used in different areas, mainly in the car industry. If we want to respect our environment, hydrogen is the future. At least, this will be so provided water is free and available to most inhabitants in the planet.

CHAPTER 5

TASK 5-1

1) Incomplete (it).
2) Incomplete (they).
3) Complete.
4) Complete.
5) Compete.
6) Incomplete (are).
7) Incomplete (it).
8) Incomplete (it).

TASK 5-2

1) Independent, independent.
2) Independent, independent.
3) Independent, independent.
4) Independent.
5) Independent, dependent, independent.
6) Independent, dependent.
7) Independent, dependent.
8) Independent, dependent.

TASK 5-3

1) Complex Sentence.
2) Compound-Complex Sentence.
3) Complex Sentence.
4) Complex Sentence.
5) Simple Sentence.

TASK 5-4

The writer uses a short, simple sentence after the long series of complex and compound sentences to make an evaluative commentary on how difficult it is to choose the appropriate engine design as regards the cycle. This simple sentence acts as the concluding sentence of the paragraph.

TASK 5-5

1-b	6-d	11-b	16-c	21-c
2-d	7-b	12-b	17-a	22-c
3-a	8-b	13-d	18-a	
4-d	9-b	14-c	19-a	
5-d	10-d	15-d	20-b	

TASK 5-6

1) This product is neither flammable nor radioactive.
2) Both stability and malleability are the properties we require.
3) Whether students like it or not, they must take an exam at the end of the semester.
4) In semi-virtual courses, home assignments can be either submitted in class or sent via the intranet. Students can choose the system they prefer.
5) It is argued that industrial injury rates depend on both economic factors and on the role of the State and safety engineers.
6) Neither unemployment nor workers' reluctance to report accidents helps to reduce workplace safety.
7) The managers interviewed are either involved in accident prevention or show high tolerance for rule violation.
8) Some foreign language learners feel that neither learning words by heart nor doing mechanical exercises provides them with enough fluency.
9) It remains to be seen whether (or not) students coming from private secondary schools outperform students from state secondary schools (or not) at university.

10) Research on composites is necessary not only in the fields of construction materials and architecture but also in the fields of biotechnology and medical surgery.

TASK 5-7

1) Since	2) but also	3) like	4) yet
5) due to	6) As a result	7) since/given that	8) However
9) as	10) as a consequence	11) Apart	12) even
13) although /even if	14) thus	15) second	16) Lastly/Finally
17) Nevertheless	18) though		

TASK 5-8

1. **Due to** their growing population, they need more food.

2. The interns in the department often do routine secretarial-type work; **for example**, they often prepare coffee and tea for their bosses.

3. The dean agrees with your suggestion for our faculty; **namely**, that smoking should be banned everywhere.

4. The function of a scientist is to know **whereas** the function of an engineer is to do.

5. He wore special clothes in the laboratory **in order not to** disobey the orders.

6/7. Banks and other businesses **such as** restaurants, bars and other shops were making the necessary arrangements in 2002 **in case** the prediction that people would prefer credit cards to Euro currency came true.

8/9. **If** many different types of gaskets or nuts and bolts are stocked, sooner or later the wrong type will be installed. It is much better to keep the number of types stocked to a minimum **so as to** minimise errors.

10. Economies of scale can generate cost advantages for large firms; **on the other hand**, important diseconomies of scale can actually increase costs if firms grow too large.

11. There are some important physical limitations to the size of some manufacturing processes; **for example**, engineers have found that cement kilns develop unstable aerodynamics above seven million barrels per year capacity.

12/13. Very specialized jobs can be very disappointing for employees. **When/Whenever** workers become mere 'cogs in a manufacturing machine' worker motivation wanes, **thus** affecting productivity and quality.

14. **Instead** of competing with Nike, Reebok, and other high-priced shoe firm, Addidas has decided to sell its basic shoes at relatively lower prices in such a highly competitive market as the US market.

TASK 5-9

1) We should send the feasibility report to our white-collar staff, **namely** female engineers aged between 25 and 35.
2) The protective ozone layer has been seriously damaged in the past decades due to the emission of chlorofluorocarbons. **That is why** humans should stop using them. Or: **Therefore**
3) It is necessary to bear in mind the unethical uses of genetics. **Otherwise**, it is impossible to understand why some governments are reluctant to regulate the scientific use of genetics. Or: **Unless** the unethical uses of genetics are born in mind, it is impossible to understand why some governments are reluctant to regulate the scientific use of genetics
4) Some raw materials will exhaust in a few years **unless** governments do something to prevent their extinction.
5) **Even though** a computer has a far bigger memory than a human being, it has to be programmed by a human. /Or: **However.**
6) Benzene is the most feared of the specific pollutants emitted by the motor **because of** its incontestable toxicity.
7) **On the one hand**, e-mail is an Internet resource which provides people with utility and ubiquity. **On the other**, e-mail requires a permanently updated security appliance to be protected against viruses. Or: **However,**
8) Skylab was to be a laboratory in orbit where men could live and work for extended periods. **However**, the initial concepts of its functions and uses evolved and changed with the passing of time.
9) The willingness to pay a premium for renewable energy and energy efficiency among the general public can be explained by demographic factors, **such as / for example** age, salary, and education.
10) The truck drive-shaft spline failed **owing to** corrosion fatigue.
11) The piece has failed **since** one of its parts has become permanently distorted and its reliability has been downgraded.
12) A special wiper system has had to be designed **so that** it cleans water and debris from the angled windshield of this aerodynamic car. Or: **which**
13) We know exactly which torque has been used on this joint. We don't know the exact preload created by that torque, **though. / Or: but**.
14) Certain metals should be protected from the rain **in case** they get rusty. Or: **Because.**
15) The structure can break **unless** the pressure is reduced. Or: **So long as** the pressure is reduced, the structure will not break. Or: **Therefore**.
16) Hewlett Packard makes a wide variety of testing and measuring instruments, ranging from $400 oscilloscopes to $500 microchip-testing systems, which require regular maintenance and calibration. **Moreover, / Consequently**, Hewlett Packard provides the maintenance and calibration service to its customers.
17) The Internet is just one technology for interacting with customers. **However**, it is the interaction itself that creates a valuable relationship between a company and its customers.

TASK 5-10

1) *Unless a special paint is used to protect the chassis*, it will be easily corroded.
2) There was a misunderstanding *as the recorded information was badly transcribed*.
3) *Although they made an enormous effort*, they still couldn't finish the project on time.
4) *In case you have lost the copy I gave you of my book*, I'll send you another one.
5) *Because the results were unexpected*, the laboratory is considering the possibility of repeating the test.
6) He telephoned *so as to find out what was wrong with the shipment*.
7) *As governments are not taking the atmospheric reduction of pollutants seriously enough*, global warming will keep on increasing.
8) *Since there was a loose connection*, the fuses blew several times.
9) *In spite of the improvement on the production method and the revision of prices*, the sales are still going down.
10) A telescope is *(a device) used to see distant bodies*.
11) *When we arrived just after 7 o'clock*, the lecture had just begun.
12) *If the other side hadn't been so far away*, they would have reached it.
13) *Unlike plastics*, metals conduct electricity.
14) *Provided you replace the ink cartridge*, you will get the desired printing quality.
15) *After the light had turned red*, the alarm rang.

TASK 5-11

Atmospheric pollution

Air is ninety percent nitrogen, oxygen, water vapour and inert gases and we breathe it because it gives us the oxygen that is essential to live. Nowadays, however, air is composed of other substances that are released into the air by human beings. (…)

The main cause of air pollution is the release of tiny particles (measuring about 2.5 microns or about .0001 inches) into the air which is mainly produced by the burning of fuel. The exhaust fumes from burning fuel in automobiles, homes and industries are the major source of atmospheric pollution. Besides, the release of other noxious gases, e.g. sulphur dioxide, carbon monoxide or nitrogen oxide, that take part in further chemical processes and reactions are also expelled to the atmosphere. **All these gases produce certain harmful effects, namely smog, acid rain, the greenhouse effect and holes in the ozone layer**.

To begin with, some chemical reactions between pollutants derived from different sources (for example, automobile exhaust and industrial emissions) cause an effect known as smog. Depending on the geographical location, temperature, and wind and weather factors, pollution is dispersed differently. (…)

Secondly, another consequence of air pollution is acid rain, which is produced when a pollutant like sulphuric acid combines with drops of water in the air and acidifies these water drops. (…)

Thirdly, the greenhouse effect is another result of air pollution, also referred to as global warming. This effect is related to the building up of carbon monoxide gas in the atmosphere. (…)

A **final** effect is the ozone layer depletion. Chemicals are expelled to the stratosphere, one of the atmospheric layers surrounding the earth; (…)

As we have seen in the paragraphs above, atmospheric pollution can affect seriously our health in different ways and, ultimately, our survival as a human race. Smog, acid rain, global warming, and the widening of the ozone layer are so pervasive that a complete eradication of atmospheric pollution is practically impossible; yet we can reduce and prevent it from growing. A thorny question arises at this stage: what actions can be taken to reduce and prevent air pollution? This is a complex and interesting question which we are not going to address here.

To sum up, nowadays steps by many governments of the world are being taken to stop the damage from air pollution (…).

TASK 5-12

1) **To begin with,** the term design is itself a modern invention, a product of the British industrial revolution, which began in the late 18th century.
2) **The other advance that paved the way for the birth of design was** the introduction of novel techniques of high-volume production.
3) **Consequently**, engineers and entrepreneurs in Britain were the first to notice and to exploit the technological and social changes that arose from this machine age revolution
4) **Having described the origins of the industrial design, let us look at** the great achievements in design of the 19th century, which were mainly big buildings and industrial facilities.
5) **Compared with the most outstanding achievements in the 19th century, those of the 20th century** proved to be far more personal, even domestic, in scale *(or at the end of previous paragraph)*.
6) **Finally, design has evolved to such an extent that in today's mass market,** every consumer can be a design critic and the more discriminating people become, the more manufacturers realize that merchandise must meet the demands made of it.
7) **To conclude,** (…).

TASK 5-13

   a) Short and long transition:
- To begin with, (paragraph 2)
- *PVC-U* and *PVC-C* are two subtypes of PVC that share some similarities but they have some distinctive features that explain why each is appropriate for different uses. (paragraph 2)
- On the other hand, (paragraph 3)
- As a result of all the characteristics described above, *PVC* is the ideal material to satisfy pipeline requirements. (paragraph 4)

   b) PVC-U and PVC-C are two subtypes of PVC that share some similarities but they have some distinctive features that explain why each is appropriate for different uses (transition sentence going over two paragraphs).

   c) *These, two, sorts, fluid pipelines, fluid channelling, subtypes, they, it, its, conductions.*

TASK 5-14

1) The two maintenance heads**, who do not get on very well at work,** enjoy meeting and playing cards at the week ends.
2) Every Christmas card is always signed by the two managers, who **consider** this time-consuming activity is worthwhile and necessary.
3) The firm negotiated over price and delivery terms **because they/ it** wanted the most reliable supplier.
4) On the leaflet, the **English Department encourages** university students to take as many English courses as possible.
5) Everyone is entitled to **his/her** pension scheme in this company.
6) The battle of the quality department **is** frequent complaints.
7) They read the words 'No mobile phones' **which were written** in red letters.
8) The driving force of these polymerizations with a sequential repeat unit **corresponds** to the formation of metal halide salts.
9) While he was writing the conclusions**, his** friend was preparing the PowerPoint slides for the oral presentation they were supposed to give.
10) Either of these cellular phones **works** well so long as you handle it with care.
11) The three packets have to be shipped back to **their** manufacturer as soon as possible.
12) If the rod glued to the casing were exposed to excessive heat, **this (the rod exposed to excessive heat)** would impair the security of the structure.
13) The company **has/have** organized a celebration to honour **its/their** three oldest employees.
14) Some people **think** that **their** career is the only important thing in life.
15) Neither his father nor his brothers **are members** of the society.
16) Remove the fine-grained sand from the liquid and analyze **the sand/the liquid**).

17) The company launched a new product into the market **which** was expected to make the company win market share.
18) I saw a hydroplane **flying in the sky**.
19) **Do** people always behave rationally and coherently?
20) Graham, **a highly reputed specialist in biocompatible implant materials**, recommended a total artificial hip replacement for the injured child.
21) The item was shipped to the client, and **nobody noticed** it was defective.

TASK 5-15

Passage 1
It is apparently difficult to estimate the cost of an R&D project in the screening process (1). In addition to the uncertainty of estimation, **it** (*subject omission*) is thought that inaccuracies also stem from deliberate under-estimations used to marshal support for a project (2). It was found that the ratio of actual to estimated costs **was** (*subject-verb agreement*) greater for project failures than for successes (3). One explanation may be the tendency not to give up a project and **continue** (*parallel structure*) to allocate funds to this project even when success seems unlikely (4). Although costs are difficult to estimate, (*fragment*) probabilities are even more difficult (5). Similarly, the project originators and **the people** (*special plural form*) with implementation responsibility **tend** (*subject-verb agreement*) to give more optimistic estimates than the average whereas those with 'a knowledge gap' about technical feasibility tend to give more pessimistic estimates (6). Therefore, estimates should be treated with extreme caution and **steps should be taken** (*parallel structure*) to minimize organizational bias on the costs and probabilities (7).

Passage 2
A quick look under the sink or in the garage of your home can reveal the number of products that **exists** (*subject-verb agreement*) to make tasks such as cleaning and polishing easier (1). During the first half of the twentieth century, Americans did not have the vast selection of consumer products **containing** (*reduced relative clause*) chemicals that we now do (2). Think about the numerous items **performing** (*reduced relative clause*) useful services for us, **such as** (*fragment*) oven cleaners, toilets cleaners, floor strippers, and the list goes on and on (3).

What is surprising **is** (*subject-verb agreement*) that many of the materials we work with or use daily are hazardous (4). As long as we use them properly, under normal circumstances, they should not pose health problems or environmental damage (5). What makes a material hazardous is **its** (*pronoun agreement*) basic chemical, physical or biological characteristics (6). If a material catches on fire easily or spontaneously under standard temperature and pressure conditions **it** (*subject omission*) is *flammable* (7). If a substance reacts violently by catching on fire, exploding, or giving off fumes when it is exposed to water, air, or low heat, it is *reactive* (8). If a material releases pressure, gas, and heat suddenly when subjected to shock, heat, or high pressure, it is *explosive* (9). Lastly, if a material emits harmful radiation, it is *radioactive* (10).

Many of the chemicals **exhibiting** (*reduced relative clause*) these hazardous characteristics are not readily *biodegradable* (11). This means that they are not able to be broken down easily into their component parts by micro-organisms (12). **These** (*pronoun agreement*) types of chemicals are said to be persistent in the environment because, once released they tend to stay intact and possess the same dangerous properties for long periods of time (13). Toxic **naturally-occurring substances** (*reduced relative clause*) can also pose problems (14). A wildlife refuge in California became contaminated by selenium, an element commonly found in high pH desert soil (15). The build-up of the selenium occurred because of the **irrigation methods used by nearby farms** (*reduced relative clause*) (16). During irrigation water carried dissolved selenium to the wildlife refuge (17). Deformities in developing embryos **that had ingested** selenium were found more frequently (*dangling modifier*) (18).

TASK 5-16

1) b   2) c   3) c   4) a   5) d   6) a   7) d   8) c   9) a   10) d

TASK 5-17

1) Smog damages not only crops but also forests and pasture lands that produce another $700 million in revenue for California each year. These natural ecosystems/They account for (…) and provide recreation (…).

2) The fundamental building block of all organisms is the cell with a composition of 85 to 90 per cent of water and a variety of chemical elements. The most important is carbon, which combines with itself to form (…) and with other elements to form (…). Carbon is the most important one. It combines both with itself to form long chains of carbon as well as with other elements to form the various organic molecules typical of living organisms.

3) Although older tanks made of bare steel failed primarily due to corrosion, new tanks, which are coated and protected from corrosion or are made of non-corrosive materials, have nearly eliminated failures induced by external corrosion. These new materials used to minimize failure began to appear in the 70s and include materials such as fibreglass.

4) Of the existing fresh water sources only some of them are usable and available in limited quantities. These precious water resources are increasingly threatened by the effects of human activities such as (...).

TASK 5-18

1) The meeting was decided to be postponed to the following week. (The actor *somebody* is of no interest)
2) (Correct. The actor is unknown)
3) Learners make many silly mistakes because of their inexperience (To emphasize the actor)
4) The books in the library haven't been organized yet (If you want to emphasize the actor the sentence is correct but if you want to avoid being offensive the passive is a better option)
5) (Correct. The actor is of no interest)
6) Expert mechanics were needed to adjust the compression mechanism. (The actor *they* is of no interest)
7) To meet a more specialized need, Dallas developed the DS1846 chip. (To emphasize the actor and to be more direct)
8) Half of the paper was devoted to explaining the method used. (The actor is redundant)
9) (Correct. Shorter and more direct than the passive form)
10) Owing to her excellent work, Sue was promoted (by her boss) to a higher position. (To emphasize the recipient)

TASK 5-19

The **designation** "FRP" is given to any number of advanced composites materials that **employ/utilize** high strength, oriented fibers that are either woven or twisted together or simply placed side by side in a lower strength polymeric matrix. In the case of most of civil engineering applications, FRP usually **denotes** a class of **materials** that uses E-glass fibers, in either a woven matt, roving, stitched **fabric**, or a combination configuration. The glass fibers are frequently **impregnated with** an isopoliester based-thermo-set matrix. As a result of this combination of composite material elements, **the manufacturing technique** most frequently **employed** in advanced civil engineering composites is pultrusion.

Let's look briefly at the pultrusion process. In this process spools of roving, strand, and **fabric** are pulled through a **device** that wets and **impregnates** the fibers with polyester resign so that when the fibers are oriented and **subsequently** pulled thorough a heated die, the thermo-set matrix is **formed**, compressed and **hardened** as a result of chemical reactions that take place in and around the fiber. The end result is a shaped FRP component that can be made to any length since what comes out of the heated die is a continuous stream of FRP material. A large saw is typically used to cut the FRP elements into discrete member lengths as **dictated** by a given design application.

TASK 5-20

1) The new manager has decided, contrary to the workers' opinion, that the profit sharing plan should be reinvested in more modern machinery.
2) The experiment will fail unless we apply a particle accelerator and higher temperature.
3) Please survey and report on the new situation / Please survey the new situation and then report on it.
4) If the pressure is kept too high for too long, the system may explode.
5) Injuries in car accidents are reduced with safety belts and air bag systems.
6) In writing, computers favour producing texts faster and more clearly.
7) I was interested in satellites when I discovered their daily life applications.
8) Although the system is working properly now, it has to be revised periodically to avoid future accidents.
9) We might incorporate more workforce into our branch in Poland.
10) The director is very much concerned about the price agreed upon not yet being changed.

TASK 5-21

1) Always make sure that car tyres have the correct pressure.
2) … a nice, white French vehicle.
3) But/However, this car is…
4) The latest technological innovations have been incorporated in cars.
5) Since the car is small, one can easily fill up the boot.
6) As the documents can't be kept in the glove compartment, they have to be put in a box.
7) I fastened the seatbelt because it has been demonstrated that fastening your seatbelt can save your life.
8) The bumper is used to reduce the effect in case of a crash.
9) Safety systems have been developed for both the interior and exterior of cars.
10) The car is known to be very low.
11) The passengers inside the vehicle….
12) The steering-wheel is one of the most important parts of a car driving system.
13) The direction is controlled by turning the steering-wheel with your hands.
14) Passengers are secured by means of the seatbelt.
15) Driving without the seatbelt is believed to be risky/a risky habit.
16) Internet has become one of the most important information technologies. That's because its possibilities are really enormous.
17) Nowadays we can enjoy the capability / ability to be connected to the whole world with our computer.

TASK 5-22

Sound cards for personal computers

A sound card is a conventional rectangular printed circuit board that is inserted into one of the free slots existing in the motherboard of a personal computer. Its main function is to convert data into an analog or continuous-time signal, which is later converted into sound by a pair of loudspeakers. In the upper surface of the card, there are two types of card interfaces that allow the user to connect other devices to the sound card; these card interfaces are the jacks and the connectors. The former could be divided into "line in" jacks, "line out" jacks, "mic in" jacks and "speaker out" jacks. The "mic in" jack, as its name indicates, allows the user to connect the microphone that is used not only for recording voice but also for giving oral instructions to the computer. The "line out" and the "speaker out" jacks perform the task of connecting external loudspeakers, which amplify the signal coming from the sound card. The latter connect the sound card to other devices such as a CD-ROM, the loudspeaker of the personal computer, or even an external joystick.

In order to understand how the sound card processes data, let's first consider which are the three main parameters that a sound card takes into account: the sampling frequency (8KHz, 22KHz and 44 KHz), the bit resolution (8 bits, 16 bits and 32 bits) and the mode of working (monophonic mode or stereophonic mode). The process begins when data is sent from a storing device to a chip in the card known as DAC (Digital to Analog Converter). This chip performs the conversion by receiving a discrete-time signal in its input and delivering a continuous-time signal to its output. The higher the sampling frequency is, the faster it works; the higher the bit-resolution is, the better the quality. The opposite takes place when recording with a microphone since now the processor has to convert an analog signal into a digital signal, and therefore an ADC is used.

TASK 5-23

Dear Ms. Jansen,

I am currently a fourth-year student at the University of Technology in Eindhoven and am interested in obtaining an internship position in the management accounting department at Shell as advertised on your web page.

Since 2001 I have been studying Industrial Engineering and Management. Most of the subjects in these studies deal with designing and optimizing all kinds of business processes like logistics, human resources and management accounting. In my last two years I specialized in the field of management account, which is the reason why I am writing to you.

During my university period I already heard and read much about Shell, but when I joined a conference last summer in London, I really became interested in your company. In fact, what appeals to me most is not only the way of working, presented at the conference, but also the fact that Shell is a big, international company.

The internship is about management accounting and takes at least six months. I could start from the first of January next year; needless to say, I am highly motivated to work for you. I am sure that my knowledge of management accounting could enable me to make a valuable contribution to your company.

I am enclosing my Curriculum Vitae. Please feel free to contact me for any additional information you may require. Thank you for your consideration.

Sincerely,

Bart Beaumont

Enc. Curriculum Vitae

TASK 5-24

Satellite uses and applications

In October 1957, when the UUSS amazed the whole world launching the first artificial satellite, *the Sputnik*, a scaring and exciting race started involving the thirst for power of the two dominant world power states. However, what was conceived at the beginning as an expression of military and technological power has turned nowadays into one of the most common tools for communication, weather prediction, object positioning or investigation. In fact, the uses of satellites are so varied that one can find applications almost everywhere.

As mentioned in the previous paragraph, satellites have many common and well-known applications, one of which is audiovisual communication. By simply placing a parabolic antenna on your roof, oriented towards the right position, you will receive the audiovisual signal of countries placed, for instance, on the other side of the Atlantic Ocean. In this case, the satellite is placed in the *geostationary orbit*, which allows a terrestrial receiver antenna to "see" always the same satellite in the same position. As a result, the receiver antenna captures the received signal as a reflection from the satellite antenna. The fact is that the satellites placed at this stationary orbit are privileged, not only because they move around the Earth at the same speed creating the sensation of a "motionless" satellite (,) but also because this orbit is one of the most limited aerospatial resources. Indeed, every country has its own section of geostationary terrestrial orbit, mostly addressed to audiovisual companies.

Satellites, just as stars centuries ago, have been found to be one of the most powerful positioning systems. Systems like *GPS* or *Galileo* provide the user with one of the most accurate estimate of position. In contrast with other satellite applications, a whole system of satellite interaction is required to provide a positioning service available all the time. This system implies expensive constellations of hundreds of satellites. Applications of this global positioning service range from the classical driving aid found in automobiles (used by drivers to find the shortest way) to airplane piloting, which is lately exploring the benefits of GPS, especially under bad weather conditions.

www.ingramcontent.com/pod-product-compliance
Lightning Source LLC
Chambersburg PA
CBHW080918220326

41598CB00034B/5611